THE IMPACT OF NEW MILITARY TECHNOLOGY

The Adelphi Library

The Impact of New Military Technology

THE ADELPHI LIBRARY 4

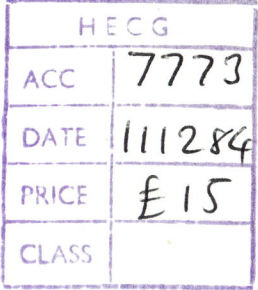
edited by

JONATHAN ALFORD

Deputy Director, IISS

Published for

THE INTERNATIONAL INSTITUTE FOR STRATEGIC STUDIES

by

Gower and ALLANHELD, OSMUN

© The International Institute for Strategic Studies 1981

Published by

Gower Publishing Company Limited,

Gower House, Croft Road, Aldershot, Hampshire GU11 3HR, England

and

Allanheld, Osmun & Co. Publishers Inc,

6 South Fullerton Avenue, Montclair, New Jersey 07042, USA

Reprinted 1984

British Library Cataloguing in Publication Data

Alford, Jonathan

The impact of new military technology −
(Adelphi Library; 4)

1. Military art and science
2. Technological innovations
I. Title   II. Series
355.02   U104

ISBN 0-566-00345-7

Library of Congress Cataloging in Publication Data

Main entry under title:

The impact of new military technology
(The Adelphi Library; 4)
Includes index

1. Strategy−Addresses, essays, lectures.
3. Arms control−Addresses, essays, lectures.
I. Alford, Jonathan. II. International Institute for
Strategic Studies. III. Series: Adelphi Library; 4.
U162.I48   355.4'3   80-67839
ISBN 0-916672-74-3

Printed and bound in Great Britain by
Biddles Ltd, Guildford and King's Lynn

0 566 00345 7 (UK)

0-916672-74-3 (US)

Contents

Introduction

Although the impact of technology on warfare is as old as the Iron Age and through history the nature of warfare (and hence the relations between states) has changed—in some cases very dramatically—as a direct result of some new invention and its application to warfare, it is a demonstrable truth that the rate of change of technology has been, and seems likely to remain, exponential. What follows inexorably from this fact is that nations that either feel threatened or have ambitions beyond their borders can never afford to allow the pace of military innovation to slow, fearing that if they once fall behind, they will never again catch up—not in quantity but in quality. And one reason why this must be so is because it is easier now to invent new things than it is to manufacture them. So long as there remains a substantial period (often up to ten years) between the inception of a new weapon system and its deployment, even the very latest weapons are out-of-date in terms of what technology could deliver. All the major powers maintain some research and development capability and the super-powers invest massively in research and development spanning the whole field of military activity, for the fears of being second, of being caught without a counter-measure, and of being outwitted in a technological sense are very great and very real. If there is political competition (perhaps expressed in ideological terms), military technology seems set to play a substantial role.

Yet there are dangers in allowing oneself to become mesmerized by technological promise. It is often much easier to make technological decisions than it is to make social or political decisions. Precisely because technology is felt not only to be neutral but to be dumb, it has often seemed to offer an easy way out. Investment in high technology has come—often mistakenly—to be seen as the solution to problems which are essentially political. The technological 'quick fix' can be seductive; it may seem to permit the reduction of military establishments by substituting high technology for manpower; and it can offer much cheaper and more powerful means of destruction. This last is graphically described in the American phrase 'more bang for a buck', most disturbingly evident in thermonuclear weapons. The debate will last for ever about whether or not technology has made the world a safer place to live in. Certainly there has been no major conflict between industrialized nations for 35 years and this may indeed be because, as Michael Howard once remarked, "the fear of the bomb is the beginning of wisdom". On the other hand the capacity for devastation which military technology has provided, while imposing great caution, makes the potential cost of error literally incalculable.

Another debate concerns the interaction of technology and doctrine: which should be paramount? Should technology serve doctrine and so produce weapons that fit preconceptions and preferences? or should doctrine be adapted to make best use of

1

what technology has to offer? These questions and others of a similar nature are addressed by Steven Canby in Adelphi Paper 109—at least in the context of Europe. He believes that revised doctrines could allow the Western Alliance to match the Warsaw Pact *conventionally*—and if that could be done with confidence, it would place the onus of using nuclear weapons first on the other side. His message is one of hope provided that NATO can rid itself of notions derived in the main from earlier applications of technology to warfare. Although he does not say so in so many words, he clearly believes that the generals are busy preparing to refight the last war despite the fact that much in warfare has changed. He is probably right to suggest that ossification has set in but in condemning it so forthrightly he may be too little sensitive to the political difficulties of a military alliance of 15 sovereign, democratic nations where progress must derive from concensus and bureaucracies are very resistant to new ideas. The pace of the slowest must sometimes seem very slow indeed, but impatience with alliance should not be allowed to blind one to the rather remarkable fact that the alliance has remained intact over thirty years. Benjamin Franklin was surely right to suggest that "we must indeed all hang together, or most assuredly, we shall all hang separately". Hanging together is bound to mean some tardiness in the acceptance of new ideas. This is one of the points that Richard Burt makes in Adelphi Paper 126 in which he outlines the debate about the introduction of new weapons technologies and suggests the direction the debate will take. "Questions will not be quickly answered" he writes and he also cautions against "the tendency to view technological developments as instant solutions to longstanding dilemmas". His paper worries a good deal about the implications for arms control of rapid technological change. There seems little doubt that the task of designing sensible and verifiable agreements to limit or reduce weapons or men becomes vastly more difficult at a time of unparalleled transformation of military inventories. All too often the wrong things have been constrained and technologists have altered the ground rules. The most obvious example of this effect is SALT 1 which constrained numbers of launchers at the very moment when MIRV technology was rendering launchers less important than the numbers and accuracy of the warheads they carried.

Accuracy is the theme of James Digby's Paper (Adelphi Paper 118) and if one had to single out only two critical technological developments in the past thirty years, the first must be thermonuclear devices but the second is almost certainly accuracy over long ranges through improved methods of guidance. Although the rather journalist conclusion that nowadays "if you can see it, you can hit it" is a gross oversimplification of the process, the fact remains that cruise missiles seem likely to land within 30 metres of a target after a flight of 3,000 kilometres, that ballistic missiles can even now be flown over intercontinental distances with a high expectation of being not more than 300 metres off their aim point and that tactical weapons can hit tanks and aircraft at long ranges and with unprecedented accuracies does constitute a revolution, the implications of which are explored by Digby.

That there is a debate about those implications is demonstrated by the two Survival pieces reprinted here. John Mearsheimer takes the view that PGMs (Precision Guided Munitions) unambiguously favour the defence against the kind of attack that NATO might expect to be launched by the Warsaw Pact. A dissenting view is given by Goure and McCormick in the rejoinder to Mearsheimer's essay entitled "PGMs: No Panacca".

Finally this volume contains two short extracts from Strategic Survey 1978. The first of these looks at how technology is altering the nature of competition in space. This points up very sharply the need to distinguish between those applications of

2

technology that can be considered stabilizing (as tending to give assured warning of attack and certainty of response) and those which can be characterized as destabilizing (such as anti-satellite warfare and the use of satellites to provide missiles with yet greater accuracy through mid-course guidance which would seem to lead inexorably to a real counterforce potential). The distinction is often difficult to see but the attempt to do so must be made for technology itself is neutral; it is the uses to which it is put that are not.

The second Strategic Survey extract looks at trends in air power—a field in which technology has traditionally flourished for the investment in machines and weapons has always—proportionately—been greater in air forces than elsewhere. The use of electronics for reconnaissance, for navigation, for weapon guidance and for computing the many complex decisions that have to be made at very high speeds is here most marked—and this demonstrates as clearly as anything can do that one very new kind of war will be electronic war—a war between beams of electrons with each striving to maintain the effectiveness of his own electronic systems while denying the other the use of his.

On 8 August 1934, Professor Lindeman (later Lord Cherwell) wrote a letter to *The Times*. In it he said: "If no protective contrivance can be found and we are reduced to a policy of reprisals, the temptation to be 'quickest on the draw' will be tremendous." Wisdom consists in applying technology to those military systems which increase the confidence that one can protect oneself against whatever threat may emerge and not in devoting it to systems which allow one to be 'quicker on the draw' than the other side.

Jonathan Alford
Deputy Director, IISS

3

1 Military Doctrine and Technology

STEVEN CANBY

INTRODUCTION

For many years now NATO has had its troubles. In the early 1960s there were long arguments over strategy and force levels, which resulted in France withdrawing her forces from the integrated military command. More recently there have been strains between the European and American partners, centred around American balance-of-payments difficulties and the connected threat of American troop reductions. As the Alliance approached its twenty-fifth anniversary it faced a distinct American reluctance to carry the same proportion of the defence burden as in the past.[1]

At the heart of most of the debates has been the problem of matching the conventional forces of the Warsaw Pact, which has in turn led to a preoccupation, particularly in Europe, with the nuclear aspects of the defence of the continent. The existence of a conventional imbalance has tended to be taken for granted and the paradox of why NATO spends more and yet obtains less conventional defence than the Warsaw Pact has gone unresolved.

This Paper seeks to address this point and in doing so to put into new perspective the relative roles of nuclear and conventional forces. It discusses how military doctrine and technology ought to interact; how revised doctrines can allow NATO to match the Warsaw Pact conventionally; and how the proper employment of new weapons can favour the defence. It suggests that, if NATO's conventional capacity is re-appraised in this way, many of the militarily induced difficulties currently besetting Alliance politics can disappear (given the will) and that a new framework for the analysis of European security problems can emerge.

I. NATO STRATEGY

Is a conventional defence desirable?
The desirability of a strategy depends upon conceptual soundness, feasibility and need, but also on its capacity to simplify issues. The old NATO strategy of massive retaliation, in use during the period when the United States had a one-sided nuclear advantage and an unambiguously credible first-strike capability, and when an American nuclear response to any Soviet aggression was viewed as iron-clad and automatic, met these criteria. On the other hand the present strategy of flexible response and graduated escalation, coupled with forward defence, has opened a veritable Pandora's Box. Flexible response by definition means discretionary action and uncertainty, and the change to this strategy made Europeans obvious and embarrassing dependants of American decision or indecision, rather than just junior but respectable partners. Giving the United States so much discretionary power generated considerable controversy. Various palliatives like the multilateral nuclear force were proposed, but the root problem was the deficiency in conventional forces and the consequent weakness for crisis management.

A declaratory strategy of flexible response without a respectable conventional capability to back it is an invitation to a *fait accompli*, as was pointed out by General Gallois and others.[2] If an opponent is told that a nuclear response is not automatic and that any

---

[1] There has, of course, been the very recent difficulty between Greece and the Alliance, but the causes of this have been regional and unconnected with the subject of the Paper.

[2] Pierre M. Gallois, 'US Strategy and the Defence of Europe', *Orbis*, Summer 1963, p. 238.

4

probes on his part will be countered by weak conventional forces while an alliance with inherently conflicting interests consults itself, he may not be deterred.[3] A *blitzkrieg* armoured attack offers the possibility of effective victory before major escalation takes place – or at a minimum the occupation of a substantial slice of German territory, after which operations could be suspended. If the conventional forces are strong, a true capability for flexible response would exist, enabling NATO to 'meet a Soviet challenge at whatever level of violence it might be presented',[1] and keep the onus of decisions on nuclear escalation with the Soviet Union. If conventional forces are weak, flexible response is a pseudo-strategy: its concept of conventional forces providing for a last warning before implementing a counterforce strategy is potentially destabilizing and is objected to by many Europeans.[5]

If NATO does not have the requisite conventional forces – and, given present trends, they are likely to become relatively even less capable unless the Soviet Union is unexpectedly generous in any mutual force reductions – what are the options? Is there a feasible alternative to flexible response? If French thinking is any clue – and France has been the most critical and thoughtful on this topic – the answer must be negative.

In the French view, the role of conventional forces is to test enemy intentions.[6] This is similar but subtly distinct from the American concept of the pause built into flexible response. Whereas the pause concept implicitly assumes no decision-maker will rationally provoke a nuclear conflict and consequently makes no distinction between hostilities arising from an unintended course of events or from deliberate aggression, the testing notion would attempt this distinction. Thus in the French concept ambiguous aggression would be countered with a conventional response, and serious aggression would be met with an automatic nuclear response, as in the old strategy of massive retaliation. This French version of a pseudo-strategy of flexible response would in theory require fewer troops, but in practice differences in troop levels are academic; they now have a life of their own and are determined by factors other than the choice of strategy. For example, the change from massive retaliation to flexible response was not met with troop increases, and France now opposes conventional reductions while the United States is a leading advocate of them (partly, one fears, because American officials are now

conveniently convinced that the conventional gap is, like the earlier missile gap, non-existent). In short, the major difference between France and the United States is the pragmatic question of who initiates the nuclear response – a remote super-power or an immediately concerned European country, like France.

Are tactical nuclear weapons an answer?
The usual way within the Alliance of attempting to overcome the problems of conventional inferiority has been to emphasize the role of tactical nuclear weapons (TNW). Europeans want them for their deterrent value, mainly because they are seen as the link with the strategic nuclear forces; but fear them because they could imply that nuclear war would be confined to Europe west of the Soviet Union. Americans, on the other hand, tend to stress their warfighting potential and the choice they provide between acceptance of a *fait accompli* and strategic nuclear retaliation. Alliance interests have clearly been different.

Recent events, in particular the possibility of American troop reductions and the escalating costs of weapons and manpower, have led to a revival of the old question of whether technology, and in particular nuclear weapons, could help to reduce defence expenditures and compensate for Warsaw Pact conventional superiority. Some argue that the use of new, small-yield, 'clean' nuclear weapons combined with high delivery accuracy could now be beneficial, because of their high firepower coupled with small collateral damage as compared with present TNW. But, while the new TNW have attractive technical characteristics, military value does not necessarily follow. To be more than just a counter and a deterrent to Soviet use of TNW, two central questions have to be answered in the positive: *Do they strengthen deterrence, and do they provide a substitute for conventional forces?*

Deterrence works in two ways: by the threat of punishment if a hostile act is initiated, or by denying the initiator the objective of his act. Strategic forces have been built around the threat of punishment; conventional forces have more to do with denial.

Deterrence by fear of punishment works differently on the two super-powers as compared with their European allies. Even though collateral damage *might* be limited, it is likely to remain a deterrent to the countries in central Europe; the two super-powers, however, risk only their field armies. The deterrent that TNW pose for the super-powers is that they may act as a fuse leading to a strategic nuclear exchange. If those who argue for the new smart TNW are correct in maintaining that nuclear warfare can be controlled and can be merely an extended form of conventional warfare, the link to strategic

[3] For a devastating critique of the pause concept, see Henry Kissinger, 'The Unsolved Problems of European Defence', *Foreign Affairs*, July 1962, pp. 521–22.
[4] *Ibid.*, p. 520.
[5] *Ibid.*
[6] Général d'Armée Aérienne M. Fourquet, 'The Role of the Forces', *Survival*, July 1969.

weapons is cut, and the deterrent value of TNW is reduced. If the contention of controllability is wrong, deterrence remains unimpaired, or even enhanced because the threshold is lower, but their use becomes exceedingly dangerous. Thus to deploy the new TNW on the battlefield to extend conventional warfare would have the effect of reducing deterrence by threat of punishment; and, if war were then to begin and tactical nuclear warfare indeed proved uncontrollable, general war would become at once more likely.

The form of deterrence the new TNW *could* increase is deterrence by denial, but this requires a relatively demanding improvement in conventional and tactical nuclear warfighting capabilities *vis-à-vis* the Warsaw Pact, at least to the point at which there is high confidence that the attacker's offensive could be stalled well forward of the Rhine. One must therefore conclude that dependence on the new family of TNW could actually worsen rather than improve deterrence. There could only be an improvement if the linkage to strategic weapons remains unimpaired and if deterrence by denial is obtained in addition. The first condition – linkage – probably does remain unimpaired, but this is inconsistent with the claimed attractiveness of TNW: that they can be used without inducing an undesirable nuclear escalation. The second condition – deterrence by denial – requires a major improvement in warfighting capabilities, and this TNW alone do not give, except in a situation where only one side of the two possesses them, which is hardly the case at the present moment.

While both sides are now armoured, the NATO may have better weapons-delivery systems, Soviet ground forces are much better structured and prepared for tactical nuclear *warfighting*.[7] NATO, because of its present deployment and style of

[7] An extensive literature developed in the late 1950s on the question of whether nuclear weapons favoured the defence or offence. Most notable in these discussions was the failure of analysts to investigate the military structure of the opposing alliances. It was commonly accepted that the United States had a ground force compatible with tactical nuclear war because the United States Army had proclaimed its Pentomic divisional organization in 1957. In reality, the reverse was true, as was belatedly recognized by the United States Army itself in its 1962 Reorganized Objective Army Division (ROAD). One problem was that the Soviet forces were mechanized, while the United States and NATO European forces remained still largely foot infantry until the early 1960s. Foot infantry lacks the tactical mobility, armour shielding, and small-arms fire necessary for the close-in protection of dispersed units. All are required for tactical nuclear warfare. Problems still remaining are that Soviet forces are better equipped for tactical nuclear warfare, and their *blitzkrieg* doctrine and organization is better suited to it than is NATO's. For an excellent summary of Soviet tactics on the nuclear battlefield, see Trevor Cliffe *Military Technology and the European Balance*, Adelphi Paper no. 89 (London: IISS, August 1972), pp. 31–34.

defence, also has the more vulnerable line of communications, supply depots and command and control centres. Under American tutelage, it has tended to think of nuclear weapons simply as greater firepower. Perhaps because of this, or as a result of their own experiences in strategic nuclear analysis, Western defence intellectuals have tended to view tactical nuclear weapons as a strategic interchange writ small, as in the well known 'shot across the bow' scenario. The Soviet Union, on the other hand, has been less bemused by theories and has concentrated upon integrating tactical nuclear weapons into a warfighting system. The Soviet forces (1) have small, highly mobile units (18 battalions per 12,000-man mechanized division and 17,000 division slice, against the United States' 12 battalions per 16,000-man mechanized division and a division slice of 48,000 or more for wartime theatre forces);[8] (2) use a

* The term division slice is used repeatedly in this Paper. A slice is the division itself plus a proportional share of non-divisional troops and administrative overheads. It is calculated by dividing force strength (total army, theatre army or field army, depending on purpose) by number of divisions. The term division-force equivalent (not used in this Paper) refers to a field-army division slice. Finally, since divisional sizes vary among countries and particularly since Soviet divisions are smaller than their Western counterparts, the author uses the term *adjusted* division slice to refer to Soviet slices that have been increased, for the purposes of comparison with those of the United States, to the same 'foxhole' strength, i.e., same number of men in infantry, tank, anti-tank and armoured cavalry/reconnaissance platoons. Within the division *itself* only 24 per cent of an American division is combat or foxhole strength in infantry, tank, anti-tank and cavalry/reconnaissance platoons. The comparable Soviet figure is 25 per cent for tank and 37 per cent for motorized rifle divisions. This partially explains why American slices (including non-divisional manoeuvre units) are only 25 per cent stronger in foxhole strength, even though the division itself is 50 per cent larger. Soviet division slices are rich in tanks and artillery, NATO's in infantry and mortars. Mechanized divisions on both sides have about the same number of major anti-tank weapons; Soviet tank divisions, however, are much less 'balanced' than Western armoured divisions and have fewer infantry and anti-tank weapons.

The peacetime United States Army slice in Europe is around 41,000. This number excludes Berlin, Southern European Task Force (SETAF), and 17,500 Army personnel in Europe not attached to United States Army, Europe (USAREUR). It does, however, include two thousand per division slice for the tactical nuclear support of allied forces, maintaining prepositioned equipment stocks for Continental United States (CONUS) based units, Special Forces and logistic support for Supreme Headquarters, Allied Powers Europe (SHAPE) – numbers that strictly speaking should not be charged against the in-theatre divisional slice.

The wartime United States Army slice (i.e. division-force equivalent) is often said to be 48,000; however, the theatre-army slice, including theatre headquarters, and various special mission troops, is considerably larger (*Reference Book: Organizational Data for the Army in the Field*, United States Army Command and General Staff College, Fort Leavenworth, Kansas, April 1972).

West European *wartime* division slices run above 40,000. Comparable Soviet slices, as adjusted upward for the same number of men in full-strength tank, infantry, armoured

linear shock tactic which is designed to hug closer to the enemy and is thus more difficult to target than NATO's more triangular 'two up and one back' deployment;[9] (3) have designed their tanks with nuclear warfare specifically in mind;[10] and (4) train under simulated nuclear conditions much more extensively than NATO forces.

As long as NATO retains its present structure, the use of TNW is unlikely to lead to favourable results in *warfighting*. New TNW and greater accuracy through terminal guidance cannot by themselves alter the fundamental structural deficiency in NATO's military posture. Improved tactical nuclear weapons, to be used against point targets rather than indiscriminately against area targets, could certainly reduce collateral damage and thus permit close-in fighting with nuclear weapons. But the very features which would make the weapons attractive also militate against their being needed as a substitute for conventional forces. The critical elements are not their nuclear capacity but the combined impact of high accuracy and timely target acquisition. Without these two features which reduce delivery error, TNW would still produce widespread collateral damage; could still not be used too close to friendly troops; and would not (without large numbers) be good tank killers because of their relatively small radius of *immediate*[11] kill against tanks (particularly with the Soviet practice of deploying armour in linear formation).

But terminal guidance enhances the performance and value of conventional weapons even more; high accuracy gives them new capabilities – in particular a tank-destroying capability for artillery and heavy mortars. Terminal guidance essentially allows conventional weapons the same capacity that it gives to TWN: close-in tank killing. More conventional weapons will be required, but they are much cheaper[12] and their lower effectiveness per round is not a serious concern if individual targets are visible.[13] This is not to say that TNW will not retain special advantages for such targets as tank units in forested assembly areas, where conventional weapons would be ineffective, but their relative advantage over conventional weapons is vastly diminished, particularly when it is considered that TNW would almost certainly have greater firing restrictions, and hence would be subject to much greater targeting delays, than would conventional weapons.

Another difficulty of the 'clean and small weapon' approach is its implicit assumption that both sides have similar tactical nuclear postures and would abide by the same rules. This is not now the case. To date the Soviet Union has emphasized large-yield indiscriminate terrain-fire weapons, many of the launchers for which are based in Soviet territory. That itself precludes the possibility of 'limiting' a tactical nuclear war. Should the Soviet Union adopt similar clean and small weapons for the battlefield, NATO's original problem of insufficient combat forces

cavalry/reconnaissance and anti-tank platoons, is 21,500. The actual manning of NATO's peacetime field-army division slices is between 25,000 and 29,000 for Germany, Britain and French forces in Germany. Reasons for the difference *vis-à-vis* the US Army vary from the disproportionately large numbers of civilian employees in the British Army of the Rhine (BAOR) to the German practice of undermanning combat units and maintaining logistic units at a sliding scale strength, whereby the most rearward units are sometimes maintained as low as 25 per cent strength. The United States Army slice also has more artillery than its allied counterparts.
[9] For an elaboration see note 34.
[10] Soviet tanks have a large number of major and minor qualitative defects compared with Western tanks. However, the area in which Soviet tanks excel is in preparation for chemical, biological and radiological (CBR) warfare in general and tactical nuclear warfare in particular. Except for filter systems, the United States and European armies have done little to protect their armour from nuclear effects. The Soviet Union, however, has: (1) designed tanks with the minimum cross section to reduce dynamic pressure damage (e.g. her T-54, T-55 and T-62 tanks are about a metre lower in silhouette than the US M-60); (2) provided radiation attenuating liners of lead and plastic, for gaseous and neutron shielding (they also secondarily provide some spalling protection); (3) provided special ventilation systems; (4) provided tanks with automatic control units to seal the vehicle against blast effects and trigger the special ventilation system.
[11] With enhanced radiation weapons, the radius in which only *delayed* kill is achieved is sometimes very large. In the long run, this may stop the attacking forces, but in the interim much of West Germany might be over-run.

[12] Cost figures for the new small nuclear weapons are classified, but various accounts in the open literature indicate an average cost of $400,000 and incremental cost of $100–200,000 for the warhead. The permissive action link alone costs $30,000. Warheads for artillery have small diameters which require expensive plutonium and are thus several times more expensive than larger-diameter missile warheads which obtain critical mass through high explosive implosion of less expensive material. In the past the greater costs of warheads, as well as the greater vulnerability of the delivery system, was more than compensated for by the dual-capability of artillery and its much greater accuracy and responsiveness as compared with nuclear rockets. New stand-off technology partially changes these considerations (see pp. 38-39). Longer-range guided missiles may now be preferable for most nuclear targets, except where a high degree of safety is required for friendly troops or maybe for targets of opportunity requiring prompt response (though the real problem is one of control procedures, not the location and type of delivery system). This suggests that most nuclear delivery vehicles could be taken from forward divisions and placed in rearward, dual-capable missile battalions. This change would not inhibit a mini-nuclear option, and it would reduce cost and increase the survivability of the theatre nuclear weapon system.
[13] As an illustration, the present United States Army division has over a hundred forward observers, and artillery and mortar tubes capable of firing a guided ('smart') shell. Since an observer can guide in a shell every 15 seconds, division artillery could *theoretically* kill every armoured vehicle in a Soviet division in one minute. If individual targets are not visible but nonetheless grouped in a known location as in an assault, mortars with infra-red seekers can have an even higher rate of kill.

would re-emerge and highlight an anomaly of the NATO defence – that of setting relatively few but high-quality, sustainable but vulnerable forces against larger Soviet numbers of smaller, less sustainable divisions equipped with cheap, relatively unsophisticated equipment.

The argument that TNW can be a substitute for conventional forces is based upon two axioms: (1) a trade-off exists between manpower and firepower, and (2) the 3 to 1 rule between the attack and the defence. The weakness of the case is that both axioms are used inappropriately. Firepower *can* often be a substitute for manpower, and if TNW were regarded as merely shells with a larger explosive yield *and* the enemy had no means of reacting with nuclear weapons there would be a substitution. But the opponent can reduce the value of this substitution by changing his tactics and can negate it entirely by responding with TNW himself, in which case the relative advantage of firepower tends to cancel out, and each side is merely likely to have high casualties. Numerous studies have indeed quite plausibly concluded that NATO's manning requirements would increase if both sides used TNW.

The 3 to 1 rule-of-thumb (that an attacker needs a numerical superiority of 3 to 1 to prevail) has led many to believe that TNW inherently favour the defender by forcing the attacker to concentrate and to become relatively more vulnerable to TNW. Neglected is the fact that the 3 to 1 generalization presumes fortified entrenchments. Hence the defender obtains his advantage at the cost of being more static and more easily targetable; if the defence is weak in combat forces, a few TNW could easily blast a gap through which armoured formations could pour. In this situation – which NATO is in – TNW are a dangerous option against a *blitzkrieg*-oriented opponent. A defender could at best stalemate such an armoured attack with sophisticated weapons delivery and real-time target acquisition. This is obviously risky, even assuming its feasibility.

In short, while NATO maintains its present military posture, the notion that TNW can substitute for conventional numbers will prove illusory. Without a viable *conventional* defence, NATO will also lack a viable tactical nuclear *warfighting* capability. Conversely, if NATO had a viable conventional defence, TNW would be unnecessary as a substitute. But this is not to deny TNW an important role. They are needed as a link in deterrence, for countering a one-side Warsaw Pact use[14] and for providing nuclear options less

than an all-out strategic response. The new family of TNW are particularly attractive in this latter category.[15] Their tailored weapon effects (accuracy, prompt radiation only, intense radiation salting, and earth penetration for cratering or shaking down structures like bridges) could be invaluable for crisis management and for a counterforce strategy. They would however be most suitable against interdiction – not battlefield – targets, indicating again their limitations as a substitute for conventional numbers.

Consequently, it must be concluded that a strategy of defence by denial based on substituting nuclear weapons for conventional numbers is illusory, and NATO's only choice is a strategy with a true capability for flexible response. The old system of equilibrium based on asymmetrical power relationships, whereby NATO deterred the Soviet Union with the offensive threat of strategic nuclear weapons and the Soviet Union deterred the United States with conventional forces and intermediate- and medium-range ballistic missiles (IRBM and MRBM) aimed at Western Europe, has been permanently disrupted. If there is to be a stable military equilibrium, NATO must be roughly equal to the Warsaw Pact in each of the three areas (strategic, tactical nuclear and conventional). NATO can no longer hope to overcome its conventional weakness by stressing strategic or tactical nuclear superiority. It can seek to overcome its conventional weaknesses in warfighting and in crisis management by emphasizing superior resolve and a superior nuclear counterforce capability, via the technological lead of the United States in flexible control of strategic weapons.[16] But the more attractive choice

[15] The present weapons are dated. Of the 7,000 warheads, about 3,000 are allocated to vulnerable forward artillery and 2,000 to aircraft still using delivery techniques with a high circular error probable. Only some 500 are assigned to rearward missile units (which admittedly have been unwieldy and either inaccurate or expensive). Of the remaining warheads, 300 are atomic demolition mines, 700 are for the obsolescent *Nike-Hercules* air-defence missile, and 400 are for land-based anti-submarine-warfare aircraft. Jeffrey Record, 'US Tactical Nuclear Weapons in Europe: 7,000 Warheads in Search of a Rationale', *Arms Control Today*, April 1974.

[16] The present American debate about strategic counterforce options is predicated on the fear that Soviet technological strides in multiple independently-targetable re-entry vehicles (MIRV) could erode the United States' present advantage and, given the limitation on launcher numbers agreed in SALT I, actually place the United States at a disadvantage in both counterforce and overall strategic capabilities. The purpose behind the options policy is to ensure that the United States has a matching capability. However, this raises a potential dilemma. Advocates argue that the credibility of the American strategic deterrent would be enhanced, but opponents can also argue that it dangerously lowers the nuclear threshold or even permits the United States to temporize against an all-out strategic response to a Warsaw Pact attack. The way out of this impasse is by recognizing that counterforce options serve both deterrent and crisis-management purposes. In the latter role, counterforce options are often viewed as

[14] As Robert Lawrence also rightly points out, the Soviet Union has (at least until recently) viewed nuclear weapons as an integral part of her *blitzkrieg* offence, not as a means of shoring up a faltering conventional capability. 'On Tactical Nuclear War', *Revue Militaire Générale*, January/February 1971, p. 257.

for NATO, which has appeared to date infeasible, is the development of strong conventional forces coupled with a strong nuclear deterrent force. Realistic suggestions for improving NATO's military posture must focus on reducing the cost of such conventional forces within the context of a true capability for flexible response, graduated deterrence and forward defence.

Is a conventional defence practicable?
While Europeans now concede the desirability of a true capability for flexible response and strong conventional forces, their practicability remains clouded by five debilitating arguments: budgetary cost and manpower, weakened deterrence, the destructiveness of conventional war, infeasibility of dual-capable ground forces, and doubts about reciprocal Soviet restraint.

The feared cost of an adequate conventional defence has been the primary argument. General Fourquet advocated conventional forces adequate only to 'test' enemy intentions; he dismissed the possibility of strong ground forces because 'the upkeep of such forces would be too heavy an economic burden for the countries of the alliance'.[17] This contention presumes that additional conventional forces would be replicas of existing ones. Matching Soviet forces on this basis would indeed require more than a doubling of present NATO ground forces in the centre region because of superior Soviet organizational and mobilization practices.[18] This would double European defence budgets and bring their income share to a point approaching that of the United States. This is clearly not an unbearable economic burden for Europe, but it is one virtually certain not to be accepted.[19]

Fortunately no such doubling of manpower and military budgets is necessary. The fact is that, even in the central region, NATO countries plus France

have more men under arms and spend more than the Warsaw Pact countries.[20] The solution is not to be found by the economic panaceas of standardization, economies of scale or division of labour,[21] nor in declaring the problem non-existent by trying to rationalize present postures, as some American analysts have done. The problem is that NATO has simply bought the wrong type of forces, just as Germany's opponents did in World War II. By grappling with the too-long-ignored problems of military production processes – *how* inputs of resource get turned into outputs of forces, what *kinds* of outputs are desired, what *kind* of tactical doctrine and organization is indicated as a consequence and *how* technology can remove certain constraints in the way forces have been organized and developed – NATO can convert its present manpower and budgetary outlays into an ability to more than match a Warsaw Pact attack.

The second argument against a stronger conventional defence has been the fear that deterrence might be weakened. A true capability for flexible response, as opposed to NATO's current posture, strengthens deterrence. As Henry Kissinger has stated, 'if a stand-still in military operations can be forced, the major objective of the defence will be achieved'.[22] A capability for stopping the enemy on the ground means for him not only little prospect of gain, but also the possibility of losing what he has.

The issue of American force reductions and weakened deterrence indicate that more than syllogistic logic surrounds NATO's strategic debate. Those advancing the weakened deterrence argument should logically support American force reductions. Since advocates of weakened deterrence contend that the Alliance's conventional forces are at best only marginally above a threshold of defensive viability, American force reductions would be more detrimental to their military than their 'hostage' content. This would favour a return to a strategy of massive and automatic retaliation or at least to the French 'testing' version of flexible response. Yet few seem willing to advocate a return to massive retaliation or

---

competitive with, rather than complementary to, conventional forces. The competitive role readily raises the spectre of nuclear warfare and the opening of a Pandora's Box. The complementary role visualizes strong conventional forces as the means of crisis management, with the counterforce capability as the measure of last resort and as the retardant to Soviet use.

[17] M. Fourquet, *op. cit.* in note 6, pp. 208–209.

[18] In terms of divisions in the central sector, NATO plus France has 29, the Warsaw Pact 58. Western divisions in general, however, have a higher state of combat readiness, qualitatively superior equipment, and about 25 per cent greater 'foxhole' strength. The Soviet Union's major advantage is her rapid mobilization and reinforcement system. Whereas NATO could reinforce its division count by 2 to 5 divisions in the first 30 days, the comparable Warsaw Pact figure is between 25 and 50 divisions. In addition many of NATO's reinforcing divisions are airborne and regular infantry, more suited for theatres other than Europe.

[19] For a comparison of European NATO and United States military expenditures, see *The Military Balance 1974–1975* (London: IISS, 1974), pp. 78, 79 and 81.

[20] In ground forces alone, NATO plus France has 1,194,000 men in central Europe as compared with 935,000 for the Warsaw Pact. NATO plus France also has another 120,000-man advantage in tactical air forces. Published statistics usually imply a Pact superiority because of the French exclusion. *Ibid*, pp. 78–79. Some statistics only include forces actually assigned to the alliances, thus excluding earmarked and other national forces.

[21] For a critique of these economic arguments, see Steven L. Canby, *Damping Nuclear Counterforce Incentives: Correcting NATO's Inferiority in Conventional Military Strength* (Santa Monica, California: Arms Control and Foreign Policy Seminar, July 1974), pp. 11–16.

[22] Kissinger, *op. cit.* in note 3, p. 530. The ability to stop an enemy must however be such as to prevent him being in possession of significant NATO territory when operations cease.

to adopt the French concept, and fewer still seem willing to contemplate matching Soviet conventional capabilities. France is among the most vociferous opponents of American force reductions. NATO's security stance thus seems neither fish nor fowl.

The third argument, that conventional war would be catastrophically destructive for Europe, is a corollary to the fear of weakened deterrence. If any war is fatal, so this argument runs, is it not better to decrease the probability of *any* conflict than perhaps encourage a greater probability of a *conventional* one? This line of reasoning assumes a long war like World War II – a highly unlikely prospect given the existence of nuclear weapons and the asymmetries between NATO and Warsaw Pact force postures. A *short* conventional conflict in Europe, regardless of its intensity for the combatant forces, would not be as destructive as in the past to civilian populations and the economic base. World War II was devastating because of its duration, the operations designed to destroy the enemy's war-making industrial capacity, the scorched-earth tactics pursued by retreating belligerents, and the use by slow-moving allied armies of massive artillery fire against reforming German defences.[23] In a short war, scorched earth tactics by retreating NATO forces would not be particularly useful in the spatial shallowness of western Europe. The wholesale destruction of industrial capacity in urban areas would make little sense, and the Soviet Union would have no incentive to attempt it if winning, and would be reluctant to divert limited conventional means to that end if losing. Military forces are now small by World War II standards. Only the United States has a large air delivery capability,[24] and only the United States now has the logistic capacity for employing massive artillery firepower. The Soviet forces formerly had large numbers of artillery divisions, but these have been sharply reduced (some, of course, being replaced by nuclear weapons). In addition, because longer-range air- and missile-delivery systems have become too expensive to employ extensively on non-military targets without tactical nuclear warheads, the depth of most of the collateral damage in a conventional war would be more or less limited to the general battlefield area (though this, of course, may move). Technological possibilities for the 1980s (discussed on pp. 30–33) may change this conclusion by providing a capability of destroying at reasonable cost a transport network as deep as several hundred kilometres from the battle lines, but this would still be only a minute fraction of the destruction in World War II and could be readily repaired with the assistance of geographically favoured allies. Moreover, such a damage capacity is asymmetric in the sense that the defence would *gain* from destroying a transport network in the enemy's rear, whereas the offence would find that destroying the defender's road complex would block his own forward movement. Finally and most important, with a true capability for flexible response, battle-lines could be held near the peacetime border.

A fourth objection to strong conventional forces is their alleged incompatibility with nuclear warfighting. As General Gallois (once seconded by Henry Kissinger[25]) put it:

It is easy to demonstrate that the two systems – the conventional use and the nuclear use – have hardly anything in common, and that in a future context, the latter must inevitably triumph over the former.[26]

General Gallois then proceeds to describe why:

Clearly, the conventional system rests on the concentration of men and material, and its firepower is proportional to troop concentration. It requires a complex logistics system capable of disposing of huge quantities of voluminous and heavy material. This form of war is dependent on time, for the 'wearing out' of the belligerents is relatively slow. There is sufficient time to make use of mobilized reserves, to step up the production of armament, and to exploit industrial resources.[27]

This statement by General Gallois describes how many believe conventional wars are fought, and NATO's military forces reflect this World War II experience. That, in a nutshell, is NATO's problem. Conventional wars do not need to be fought in this

---

[23] The early German victories caused little civil destruction, though the newspaper headlines of the time may convey a different impression.

[24] Soviet air forces do not have a large carrying (range times payload) capability. Their tactical aircraft were designed with the emphasis on air defence; the bulk of these fighters in a short-radius (100 miles) high-low-high configuration have a payload of only two 550 lb bombs. The remainder of the Pact's fighters – about 1,500 Su-7 and MiG-21J and K – can carry two 1,100 lb bombs 300 to 500 miles in a maximum-radius high-low-high penetration mode. (An F-4 by comparison can carry about 11,000 lb for a similar delivery mode.) Recent Soviet aircraft developments like the Su-17, Su-19 and the MiG-23 have led to speculation (and military service rationalization) that the Soviet Union may be changing her traditional tactical air doctrines and moving towards a posture more akin to that of NATO. While the Su-17 is a modification of an earlier basic Su-7 design, the Su-19 and MiG-23 are dramatic departures from the standard design philosophy and point towards NATO's larger, more complex, more versatile and more costly aircraft. The Su-19 in particular is the first aircraft the Soviet Union has designed for the close air-support mission since World War II. In the past the Soviet air forces have been mainly oriented towards neutralizing Western air power; their strike power has, unlike NATO, been concentrated in their ground forces.

[25] Henry Kissinger, 'Limited War: Conventional or Nuclear?' *Daedalus*, Fall 1960, p. 811.
[26] Gallois, *op. cit.* in note 2, p. 246.
[27] *Ibid.*

fashion. Certainly the Soviet Union does not appear to contemplate fighting in this way.[28] Furthermore, as will be argued in this Paper, *defensive* tactics suitable for *armoured* warfare and tactical nuclear warfare can be quite compatible. But it is true, as General Gallois really indicates above, that the (Western) tactics of World War II warfare and tactical nuclear warfare are incompatible.

A fifth objection to strong conventional forces is that, since the Soviet forces have always envisaged a conflict in Europe as nuclear, strong conventional forces are unnecessary and wasteful. While Soviet doctrine has moderated this position in recent years, this argument in any case misses the essence of conventional forces in a graduated strategy such as flexible response. Formerly, conventional forces were the *means* by which a state defended itself. In a conflict among nuclear powers, conventional forces no longer have such status: they are now there to meet lesser threats and as a *means* for crisis management (what Thomas Schelling calls strategic-political manoeuvres in a competition of risk-taking[29]). Accordingly, a nuclear power without comparable conventional forces is at a disadvantage in a confrontation with another comparable nuclear power. Hence the purpose of strong conventional forces in Europe is no longer really for warfighting but to prevent any Soviet advantages in 'brinkmanship' and crisis management.

Still another purpose of strong conventional forces is to deny the self-confirming aspect that any war in Europe will become nuclear. One can agree that a strategy of flexible response and an initial limitation to the 'lower' and less destructive forms of warfare require a mutual restraint that the Soviet Union believes is too contrived for Europe (which in itself shows the Soviet pre-occupation with deterrence in Europe and is indicative of non-aggressive intentions). But one can also agree that the current asymmetries in strategic and conventional force postures would be most likely to induce instability and early escalation to nuclear weapons.[30] Since Soviet conventional forces are organized to overrun Western Europe quickly in a short war, it is highly probable that either NATO (as structured at present) would feel the pressure to opt for tactical nuclear weapons (to inhibit the Soviet onslaught) or the Soviet Union would pre-empt. Strong conventional forces and a true capability for flexible response, as Henry Kissinger has termed it, would reduce these

[28] See for instance Trevor Cliffe, *op. cit.* in note 7, pp. 29–35.
[29] Thomas C. Schelling, 'Nuclears, NATO, and the "New Strategy"', in Henry A. Kissinger (ed) *Problems of National Strategy* (New York: Praeger, 1965,) p. 173.
[30] See for instance, Steven Canby, *NATO Military Policy: Obtaining Conventional Comparability with the Warsaw Pact*, R-1088/ARPA (Santa Monica: Rand Corporation, June 1973), pp. 81–86.

pressures and raise the nuclear threshold to both sides' advantage.

In short, NATO's problems with flexible response stem not from an inherent lack of logic, but from improper implementation. The pseudo-strategies of flexible response calling for a pause or testing are strategies assuming military inferiority. They can only aggravate the Alliance's internal difficulties; they can never solve them. A true strategy of flexible response, now that the Soviet Union has obtained a measure of strategic parity, holds meaning only so far as NATO can produce a viable conventional defence against the concentrated armoured forces of the Warsaw Pact. Indeed, a strategy assuming military balance among the three categories of strategic, tactical nuclear and conventional warfare is crucial if central Europe is to be defended and the nuclear threshold avoided (and for that matter if containment of the Soviet Union is elsewhere to be achieved). By preparing itself for the sustained, long-term, infantry-style fighting of the last war, NATO may not survive to respond – flexibly or otherwise – after the opening battles of any future war.

Restructuring

Thus, NATO's forces must be restructured to counter the kind of war that its potential adversary is contemplating and to resolve the basic paradox of military wealth and combat inferiority. The restructuring alternative rejects the view that NATO must either increase its military budget or resign itself to a conventional inferiority. This inferiority exists because its military doctrine and its derived organization are no longer appropriate to the present European context. The structure and operating practices of the forces must be adjusted to fit present conditions. In particular, the 'all-purpose' British and American organizations that have influenced European military thinking (and which were themselves influenced by pre-war French military thought) are not sufficiently attuned to existing realities in Europe. They are oriented towards infantry warfare, not a rapid armoured attack such as the Warsaw Pact is prepared to execute. Standardized equipment, for example, suitable for geographic and climatic extremes where the United States with her worldwide obligations must be prepared to fight, is unnecessarily expensive to procure and to maintain for the requirements of western and central Europe. Nor is an expeditionary organization required, whereby the force can be organically self-supporting for an indefinite period. Forces deployed in Europe do not need to be logistically balanced because local resources can be tapped and *armoured warfare does not require the logistical balance that an infantry-style, across-the-front deployment requires*. In addition short conflicts require less maintenance (equipment

11

can be permitted to run down), fewer engineers (road and rail networks will suffer less destruction), and in general far fewer logistic troops than longer conflicts. A military force designed logistically for a short war can usually also fight a long war if necessary, provided long lead-time items are hedged and economic resources are available. But a military force which is structured for a long war – and which is consequently weak in short war capabilities – may not survive to marshal its potentially superior power.

The Soviet forces, on the other hand, while often equipped with older and technically inferior weapons, have developed a superior mechanism for waging the only type of conventional war they could expect to win against NATO – a short, intensive campaign that forecloses NATO's ability to mobilize its superior overall resources over time. They have overcome technological inferiority by quantitative superiority in *combat* units and by tactics which minimize the need for sophisticated equipment. They have streamlined their formations by recognizing the economies inherent in *blitzkreig* warfare and their own special style of steamroller tactics. Finally, while NATO armies have tended towards an American-type reserve and replacement system, the Soviet Union has retained and modified for armoured warfare the German-developed mobilization system designed to flesh out rapidly a large army and to surge over her opponent in a series of quick campaigns.[31]

Essentially, the Soviet Union has no alternative to her plan for a *blitzkrieg* type of war, and this is the kind of attack that NATO must be prepared to repel.[32] As the defender, NATO should be able to buy ade-

quate forces at less expense than the Pact because it need not project offensive military power other than in counter-attacks within a basically defensive strategy, and much new technology is *inherently* more advantageous to the defence than the offence. For example, a defensive force has a target-acquisition advantage and can rely extensively on pre-positioned forward supplies and mobilized civilian assets, while an attacking force must project itself and needs a more developed logistic system to maintain its more actively used combat equipment, to move its supplies forward and to repair a transport network damaged by a retreating enemy.

Because the Soviet Union has designed her forces to peak early, and for pencil-like armoured thrusts, a countering-force design would enable NATO to match the Pact's initial combat capability without creating long-run vulnerability or sacrificing superiority in mobilization potential. In short, exchanging the current NATO long-run capability, designed for across-the-front offensives, for larger numbers of streamlined, more specialized division-equivalent forces and the ability to integrate reserves rapidly into the active forces would have an optimizing effect: greater deterrence, greater war-fighting capabilities, a higher nuclear threshold and major cost savings from simpler equipment and sharp reductions in military components no longer central to the task of immobilizing forces on the battlefield. The remainder of this Paper, after a brief look at how the Soviet Union uses her resources, is devoted to an analysis of the ways in which NATO forces might be reorganized to achieve such aims.

II. SOVIET WARFIGHTING CONCEPTS

Military power from an inferior resource base

Although the Warsaw Pact possesses more short-term mobilizable strength than NATO, it has far smaller long-term resources. If this is the case, why is it generally thought impracticable to match the Pact's conventional power? The answer to this apparent paradox is organizational; the side inferior in resources cannot fight *on the same terms* as its superior opponent and expect to win. It must create asymmetries in force designs which give it the appearance of strength to reinforce its diplomacy and the reality of strength to meet military contingencies, which it will hope to control by taking the initiative. It must therefore seek to: (1) develop a superior

military system, as did Germany with her *panzer* armies in World War II, or (2) develop a more rapid mobilization system in order to defeat the enemy before he can organize his superior resources, as Germany attempted in World War I. Thus the Soviet Union has been attracted to a strategy that precludes the development of NATO's much greater military potential: a quick and overwhelming offensive developed from organizing specifically for armoured warfare and from rapidly deployed reserves. At the operational or grand tactics level, the Soviet Union has adopted and improved the German *blitzkrieg* concept; at the tactical level, she has retained a number of practices based on her own tradition of mass attack. The ensuing organizational economies and a cadre mobilization system have led to a large, rapidly deployable force structure.

Soviet postwar military aims in Europe have been designed to offset American strategic power (while Soviet strategic strength was built up) and maintain

[31] For an account of its origins and successes, see William O. Shanahan, *Prussian Military Reforms, 1786–1813* (New York: Columbia University Press, 1945).
[32] For a catalogue of reasons why the Soviet Union would be disinterested in a protracted conventional conflict in Europe, see S. Canby, *op. cit.* in note 30, pp. 81–86.

control of Eastern Europe. While Western Europe rebuilt its strength under the American security umbrella, the Soviet Union managed to rebuild an even more devastated economy and to retain and further develop a conventional force capable of holding Western Europe 'hostage' as an indirect counter to the American nuclear threat. This land power coincidentally resolved the potential Eastern European problems. The question that has never been fully explored is *how* the Soviet Union could create such a force with fewer resources and a smaller peacetime military budget.

Both aims were achieved by maximizing force visibility, accomplished by minimizing *logistic* support so that most resources could be channelled into *combat* units. The high ratio of combat strength to support was obtained in four ways: (1) centralizing logistic support; (2) integrating much training into the operational structure; (3) organizing into smaller units than the Western counterparts, at all levels; and (4) maintaining cadre divisions in the active force. The effect of these steps can be readily compared by simply dividing total army strength by total divisions: the United States has 60,000 men per peacetime division slice; the Soviet Union 11,000. That the Soviet Union does not spend much on logistics is also apparent from comparing deployed forces. *Adjusted division slices for the peacetime field armies in Germany total approximately 41,000 for American forces but only 21,500 for Soviet forces.*

Organizational implications of Soviet military doctrine

The Soviet doctrinal linchpin is the *blitzkrieg*, the concept of overwhelming an opponent quickly through that attack. The tactic is to concentrate on narrow sectors of the front in order to break through the defences and then pour into the rear areas, enveloping the main forces and paralysing any reaction. If only one or two breakthroughs are required because of the weakness of the defence or lack of depth, the enemy can be defeated in a matter of weeks. Against stronger defences a series of breakthroughs may be required before the enemy collapses.

An important implication of these tactics is their effect on logistic requirements – an insight that the Soviet forces exclusively seem to recognize. First, if a breakthrough is decisive against a relatively weak opponent who lacks depth to his position, the war will be over before the armoured breakthrough force needs extensive maintenance or replacement. NATO is potentially in this weak situation because with a rate of advance of perhaps 100 km per day, as Soviet doctrine envisages, the Rhine and the Netherlands are only one to three days' march from the West German border in the north and centre. In such

circumstances, a large logistic tail organic to NATO divisions (and logistics in general) is not only an unnecessary expense, diverting resources from combat units, but it also restricts mobility by requiring protection and clogging road space.

Second, *blitzkrieg* tactics use armoured forces in such a way as to require relatively little logistic and indirect fire support. Large-scale fire and logistic support are required only during the heavy fighting of the initial breakthrough phase.[33] Once a breakthrough occurs, an armoured penetration exploiting enemy disorganization requires relatively little artillery and logistic support. To provide all divisions and corps (usually called armies in the Soviet lexicon) with their own logistic capability for such infrequent occurrences as the breakthrough phase would be inefficient.

The Soviet solution was an adaptation from the steamroller tactics of World War II.[34] Divisions designed to swamp enemy defences do not need an elaborate infrastructure. Following their *debacles* in 1941–42, Soviet divisions were stripped of skilled specialities, which were concentrated at higher levels. Soviet forces have thus grown accustomed to sparse support, but the innovation was in recognizing that this was a natural complement to the *blitzkrieg*. Since high-intensity combat would only occur along narrow sectors of a wide front, higher headquarters (in this case the Soviet front) can concentrate logistics to provide a support framework in which combat divisions and even armies can be used like drill tips on a high-speed drill – to be ground down and

---

[33] This assumes that the enemy has strong forward defences (which NATO does not have in central Europe) and that tactical nuclear weapons are not used to open a breakthrough gap.

[34] Steamroller tactics, at the divisional level, are characterized by a relatively inflexible command system and a rigid system of echeloned forces with few intermediate reserves (except for anti-tank). As formations are exhausted by fighting they are replaced rapidly by other echelons behind them, instead of being replenished and reinforced by fresh men or units as is Western practice. By maintaining momentum with large numbers of formations, Soviet forces plan to saturate enemy defences and offset the need for flexibility and initiative at the company level, where their tactics tend to be rigid. Having large numbers available gives higher commanders considerable flexibility.

Another characteristic is the very high percentage of divisional combat platoons in actual contact with defending forces (30 per cent as compared with 18 per cent in United States Army doctrine). This is the result of attacking with platoons abreast within companies, rather than the Western practice of two forward and one behind, and of substituting anti-tank units for tank and infantry strength held in battalion and regimental reserve. This tactic gives Soviet divisions considerable shockpower to break through defences and to create the illusion of strength in secondary sectors. It also reduces the vulnerability to nuclear weapons (as compared with NATO's normal triangular force deployment).

13

replaced until penetration occurs.[35] After penetration the divisions rapidly exploiting their advantage no longer require extensive support.[36]

Centralizing logistic assets under higher commanders and allocating them according to the demand for services is attractive, because (1) the Soviet Union has the resources to create more divisions than can conveniently be fought simultaneously; (2) divisions in combat require differing amounts of support according to the task being performed; and (3) losses among support troops are always much smaller than among tank and infantry soldiers.[37] This means that logistic *im*balance is feasible, and more combat units can be created than can be logistically supported in combat simultaneously. Thus, once a threshold level of divisions is reached, the possibility arises of stripping support from divisions and building support frameworks at higher command levels from which a large number of unit replacement divisions can operate.

Mobilization and reinforcement
The 167 Soviet divisions are at widely different categories of readiness, roughly measured in terms of unit strength and equipment levels. Some, such as those in the Group of Soviet Forces, Germany are at near full manpower strength, others apparently as low as 25 per cent.[38] The common and key element is the cadre which provides a command structure and a functioning team that can be rapidly fleshed out on mobilization. Specialized combat and other equipment is largely stored to reduce maintenance, to maximize readiness and to minimize wear and tear.[39] These features are crucial in the Soviet system not only because of the lean logistics but also in order that older equipment can still be used for lower-readiness Soviet divisions or for allies.

Planning envisages divisions becoming operationally ready at intervals according to their peacetime manning levels. This graduated readiness meshes with the concept of fighting and the limitations of the transportation system. Soviet forces accept the inevitability that, for the sake of speed and other advantages, many of their attacking units may suffer heavy casualties, and the large number of divisions mobilized is to provide phased replacement divisions so that the momentum of the attack can be maintained.

III. MILITARY DOCTRINE AND TECHNOLOGY

How technology has been used
Nuclear weapons have had many effects upon military planning but one effect that has not been appreciated is their distortion of the study of warfare itself. Strategic nuclear analysis is different from the analysis of more traditional warfare – a fact that American experience in Vietnam, a campaign conducted by two Administrations steeped in

analytical tools of strategic warfare, should indicate. Defence intellectuals have claimed that generals did not understand modern (i.e. nuclear) war; but it can also be said that academics have not understood the lesser forms of war. At root, strategic analysis is built on the concepts of assured destruction and damage limitation and is a study of the multiplied probabilities of a missile exchange and their impact upon 'rational' decision-makers (defined as those with an economic calculus). Conventional analysis, on the other hand, is not so easily structured and requires a detailed and historical understanding of military doctrinal and organizational patterns – the very type of *institutional* analysis unfashionable in academia.

---

[35] The Soviet forces now train extensively at night in order to be able to fight a continuous 24-hour battle, keep up round-the-clock pressure against the enemy – the object being to exhaust NATO units physically while theirs are replaced and the process repeated.

[36] This does not mean that the streamlined divisions are incapable of heavy fighting without support from higher headquarters. The Soviet forces hedge against this possibility by providing divisions with enough organic support until they can be reinforced from higher headquarters. This phenomenon is best evidenced by the high ratio of artillery tubes and cheap multiple rocket launchers organic to divisions. This facilitates fire-support co-ordination with manoeuvre elements, though the fire cannot be sustained without higher level resupply support.

[37] For instance, in World War II, US Army casualty rates were ten times higher in the infantry, armoured and armoured cavalry than the average for the other branches. Even within a division, a rifleman was, respectively, 15·4 and 24 times more likely to become a casualty than a divisional truck driver and vehicle mechanic. G. W. Beebe and M. E. De Bakey, *Battle Casualties* (Springfield, Ill.: Charles F. Thomas, 1952), pp. 38 and 42.

---

[38] Soviet divisions have three degrees of combat readiness: Category I, between three-quarters and full strength, with complete equipment; Category II, between half and three-quarters strength, with complete fighting vehicles; and Category III, about one-quarter to one-third strength, possibly with complete fighting vehicles (see *The Military Balance 1974–1975, op. cit.* in note 19, p. 9).

[39] *Combat Service Support*, US FM-100-10, Department of the Army, October 1968, p. 10–11, states: 'Proper maintenance of equipment increases its period of economical usefulness, reduces supply requirements and conserves resources for other purposes.' Thus while the United States Army stresses maintenance to obtain these three objectives, the Soviet Army stresses *training substitutes* so equipment *does not have to be used.*

In strategic warfare the technology is complex but the implications of new technological developments are more readily apparent because of the saliency of the equipment and missile-exchange concept. In conventional warfare the technology is often simple, but the diversity and interaction of the many systems required to accomplish specific tasks often masks the implications of new technological developments. In strategic warfare a new technology like multiple warheads can dramatically alter calculations and indeed the whole milieu. In conventional warfare a better tank or aircraft may give a temporary advantage, altering accustomed exchange ratios of enemy to friendly losses, but the opponent soon learns to cope with the technical change. Most new weapons accomplish even less, simply improving upon their predecessor in some proxy indicator of effectiveness - such as weight, speed, payload or reliability. Unfortunately many measures of effectiveness have become almost irrelevant to land combat – the classic example, of course, being jet aircraft speed.

In the United States, technology has often been seen as a panacea. If the enemy has superior numbers, the apparent solution is to offset quantity with quality. If the war is not going well, as in Vietnam, the scientist is called in to design some new gadget to make 'experience proven' doctrine workable. NATO has similarly sought technological solutions to its conventional inferiority, but none has yet appeared that has significantly improved NATO's relative position. Clearly then, the full benefits from new technologies have not been materializing.

The three most significant (and symptomatic) technologies are tactical nuclear weapons, the helicopter and precision-guided munitions. NATO introduced tactical nuclear weapons to compensate for its conventional inferiority, but, when the Soviet Union armed herself with these weapons a few years later, it was belatedly recognized that a two-sided use spelled no advantage and might even be a disadvantage for the defence because of the ease with which the offence could blast a gap through linear defences.

The helicopter has found strong advocates for its battlefield successes in Vietnam. Judiciously used, the helicopter is valuable. Unfortunately, in Vietnam its utility as a means of rapid transport led to battlefield but not overall successes. It can indeed be argued that the helicopter may have been instrumental in inducing the United States Army to rely upon its material advantage and opt for a costly but self-defeating operational strategy. Instead of solving the problem of insufficient and inappropriately organized infantry, the Army sought rapid mobility to move what few infantry it did have.[40] Instead of

focusing upon the enemy's tenuous infrastructure, which gave sustenance and regenerative capacity, the Army sought to fragment the enemy into little units by regressing through Mao's classic three-stage build-up through main-force operations by conventional forces.[41] This strategy unfortunately led to heavy American casualties and also to high collateral damage in an attempt to hold casualties down.

For European warfare the helicopter has undisputed uses for command reconnaissance, cargo and ambulance lift and a variety of administration tasks. Its employment on any scale for air-lifting troops is another matter, as also is its use in the anti-tank role.

There are many specialized combat tasks for which helicopters could be useful: carrying infantry parties for small-scale raiding; quick reinforcement of threatened areas or of units needing an infantry supplement urgently; carrying small anti-tank or road-blocking groups. Their use on a large scale is not likely to be cost-effective. Helicopters are expensive to buy and maintain, and the forward air environment could be lethal for them.

The mobility they provide can compensate in some degree for a lack of combat reserves, but the solution to this particular problem lies in finding more reserves rather than concentrating on providing mobility for fewer men.

The use of the helicopter as an anti-tank platform is also open to question. Some tests have shown the helicopter to have a high tank-destroying capability, though the scenarios used may well have put the tank at a disadvantage. In all probability the tank could soon adapt itself (albeit at some expense), from which it could be concluded that a limited deployment of anti-tank helicopters might be cost-effective, since the enemy must either face high adaption costs or an adverse kill-ratio from a limited threat. The most damaging arguments against the anti-tank helicopter are that it is basically only a substitute for tactical airpower and that it will have difficulty surviving in the context in which it is most needed – that of decisive armoured penetration where the line of contact will often not be known. Moreover if structural changes are made in air forces – as discussed later – to reflect tactical changes in the deployment and capabilities of ground forces, tactical air power has significant advantages over the helicopter in such features as ruggedness, payload, range and greater stand-off distance for weapons delivery.

The latest of the significant new technologies are precision-guided munitions (PGM), popularly known as the 'smart bomb'. As with tactical nuclear

---

[40] During the peak period of involvement, the infantry content (i.e. men in infantry platoons) of the forces in South-East Asia was only 7 per cent of theatre strength and consumed only 2 per cent of the war's cost.

[41] For an explicit exposition of this thesis, see General Richard G. Stillwell, 'Evolution in Tactics, the Vietnam Experience', *Army*, February 1970, pp. 14–23.

weapons, PGM have been acclaimed as one way to compensate for NATO's inferiority. Because of certain inherent advantages that the defence has, in target acquisition and in fighting from field fortifications often screened by woods and forests, PGM ought to work in favour of the defender. Yet, because of NATO's force structure, symmetrical adoption of PGM – inevitable one day, as the Soviet Union develops the necessary technology – may well work to NATO's disadvantage:[42] NATO's military philosophy has called for high investment-low attrition forces; Soviet military thought has opted for low investment-high attrition forces. They therefore have more targets to be destroyed and more means of delivery (aircraft and artillery),[43] a combination with ominous implications.

How technology should be used

This brief review of the three most significant technologies strongly suggests that the real dividends from new technology are not being obtained, largely because it is being used to improve existing operating

practices on the margin rather than seeking new ways of applying military force in which technology can be used to its full advantage. New technology should be viewed as releasing the constraints upon present operating practices, rather than, as now, being constrained by them. The rewards from new technology will not come from one-to-one substitutions of better weapons for old ones or in shoring up obsolescent doctrines (which should alter as the underlying technology fades), but in recognizing the implications of technology for doctrine and organization.

NATO's thinking on technology has been too dominated by the quantity–quality trade-off: the idea that numerical inferiority in *combat* forces can be offset by technological superiority. Against technologically less advanced countries unable to produce a sophisticated warfighting machine, off-setting quantitative inferiority with technological superiority has been an historically attractive proposition. Against countries with similar technologies, of course, this 'solution' does become questionable.

Empirical evidence suggests that, charted over its life span, a technology tends towards an S shape: increasing performance for cost in its early growth phase, falling returns during its maturing phases.[44] Thus as NATO precedes the Warsaw Pact along the common S curve, NATO's equipment costs will increase faster than those of the Warsaw Pact if it desires to maintain a constant level of technological superiority. This implies that if both sides rely on similar technologies with similar rates of advance, NATO can maintain qualitative superiority only by increasing budgetary allocations or by diverting funds from force structures to technology. However, increasing military budgets are no longer politically acceptable in most Western democracies, and reducing force structures would make any existing combat-inferiority gap greater and feed a demand for still greater technological superiority. Obviously, then, trying to offset quantitative force inferiority solely by exploiting new technology is a policy marked by constantly diminishing returns.

A technological solution to the imbalance is achievable only if the side in the technological lead can improve the performance of its forces at costs equal to or less than those incurred by its opponent in attempting to catch up. But because innovative technology always costs more than imitative technology, and because the cost of improving any aspect of technology at a steady rate increases with time (marginal improvements in a mature technology tend to cost more and take longer than major

---

[41] For an exposition, see S. Canby, *Regaining a Conventional Military Balance in Europe: Precision Guided Munitions and Immobilizing the Tank*, WN-8222-ARPA (Santa Monica: Rand Corporation, August 1973).

[42] In his widely quoted analysis, A. Enthoven indicated that NATO and the Warsaw Pact are roughly equal in artillery and mortar tubes. This was misleading. It implied equality in indirect firepower, when in fact the Soviet Union had an edge in artillery tubes and a considerable advantage in saturation-fire multiple rocket launchers. NATO's advantage was in infantry light mortars. In any case, equality in total tubes no longer exists. When Enthoven first presented his thesis in 1968 (*Report of the United States Delegation to the Fourteenth Meeting of Members of Parliament from the North Atlantic Assembly Countries*, United States Senate, 91st Congress, 1st Session, pp. 50–59), the Soviet forces were in the process of significantly strengthening their conventional artillery and upgrading the calibre of their mortars. Soviet divisions, though small, now average roughly the same number of artillery tubes as American divisions (*The Military Balance 1974–1975, op. cit.*, p. 83). The Soviet divisional slice in East Germany also has as many artillery tubes as the American slice and more than other NATO slices. Because of the large numbers of Warsaw Pact divisions, the Pact now has considerably more tubes than NATO.

Enthoven also argued that NATO had significant advantage in air power: numerical inferiority in aircraft was more than offset by qualitative superiority in payload, loitering time and crew training. (Alain C. Enthoven and K. Wayne Smith, *How much is Enough?* New York: Harper and Row, 1971, pp. 154–156). Symmetrical use of PGM would redress this equation in favour of the Soviet Union. Laser and electro-optical guidance simplifies pilot training. Accuracy means smaller warheads to kill point-targets and a potential ability to carry more weapons; but this works to the advantage of the Soviet Union because ordnance air resistance and the limited number of so-called aircraft hard points prevent the carriage of a full notional payload of 'smart' ordnance. Pilots also have a psychological limitation as to the number of passes per sortie each can make in a high-attrition air defence environment. With PGM each pass means one shot rather than unloading one's ordnance in a salvo or ripple.

[44] Erich Jantsch, *Technological Forecasting in Perspective* (Paris: Organization for Economic Co-operation and Development, 1967).

improvements in a new technology), technological solutions to force imbalances are practical only if revolutionary technologies can be periodically introduced.

Even if new technologies become periodically available for exploitation, an important caveat must be recognized: the innovative technology will need to be properly applied by military institutions which in many instances are composed of elements committed by tradition and instinct to preserving their expertise in familiar, experience-proven areas. Without institutional adaptiveness, potential technological superiority can be meaningless. The technologically inferior opponent may be more innovative, in practice, because of greater readiness to accept doctrinal innovations and institutional changes.

If military planning is not made with intellectual vigour so as to challenge existing dogmas and to produce innovations, operational strategies and weapons will be less than optimal. For instance, in Vietnam the United States selected a military approach based upon mobility and firepower, yet these expensive measures, while important, were not central to a strategy for counter-insurgency – which must protect the people from the insurgent infrastructure and separate the insurgent from his source of sustenance.[45]

No prescription exists for ensuring innovation; we can only analyse how we might like to organize and deploy, given technological possibilities. The goal is what the noted economist Joseph Schumpeter termed 'dynamic destruction' of replacing the old with the new. This process requires that the military art be learnt, but it also needs an understanding of how and why doctrines become obsolescent. As with society, it is ideas (of whatever validity) that drive the military: its tactics, deployments, operating codes, organization, and even its research, development and procurement choices.

In Europe the central problem is to stop the enemy enjoying tactical mobility. If this can be done warfare will become stalemated, to the advantage of the defence, as in World War I. Conventional warfare then reduces itself to (1) the possibility of seizing terrain before the defender organizes himself,

and (2) firepower exchanges. The latter, however, has lost most of its allure. The theory of coercing an opponent through the threat of large-scale destruction might have had some validity (though the empirical evidence from bombing Britain, Germany and North Vietnam would seem to refute it). But, with the overhanging presence of nuclear weapons, conventional weapons cannot be expected to have great coercive power. Conventional war should therefore be concerned with thwarting the attacker and preventing his forces from occupying ground. Combat and support forces (and their underlying technology) which are not geared to this objective or are needlessly expensive are redundant.

The specific problem is how to design a defence to counter Soviet tank tactics and to redress the attacker's advantage over the defence. New weapons introduced in tandem with a fundamental organizational change *can* lead to an historic cross-over in the age-old cycle between offence and defence. In the past the cross-over has normally been associated with a *simple* invention: the stirrup made cavalry superior to the infantry; the hook on the end of a spear and the English long bow were too much for the iron knight. Today's problem is the tank: in conjunction with tactical air power it restored mobility and gave advantage to the offence after the machine gun and artillery shell stopped movement in World War I. Henry Kissinger certainly erred when he said that:

Conventional warfare favours the defence. It has been truly remarked that but for the development of nuclear weapons, the defence would long since have achieved ascendency over the offence. Even in World War II, the attacker generally required a superiority of three to one.[46]

As in World War II, the difficulty for NATO *today* is that while the defence may well have a local advantage of three to one, it could rarely concentrate enough anti-tank capability to offset the offence's even greater advantages of the initiative and concentration of force for an armoured breakthrough.[47]

---

[45] For the thesis that a firepower and mobility orientation is appropriate for a technological power like the United States, see Colonels Zeb Bradford and Frederick Brown, *The US Army in Transition* (Beverley Hills and London: Sage Publications 1973), pp. 64–69. For the counter-thesis of specifically designing American forces to cope with insurgency and its supporting infrastructure, see Steven Canby, Brian Jenkins and Richard Rainey, *An Alternative Strategy for Vietnam*, D-17262-ISA, (Santa Monica: Rand Corporation, September 1968); and S. Canby, *Comments on 'An Alternative Strategy for Vietnam': A Theory of Victory, Quadrillage, Reserve Strengths, etc.*, D-18155-ISA, (Santa Monica: Rand Corporation, November 1968).

[46] Henry Kissinger, *op cit.* in note 25, p. 809.
[47] An inability to concentrate anti-tank weapons (ATW) has been a particular NATO deficiency. Apart from recent German Army changes, NATO has not had specialized anti-tank units above platoon level – as opposed to the Soviet practice of organizing anti-tank companies, battalions and regiments. NATO's infantry until recently has also been weak in the quantity *and* quality of its ATW, requiring tanks to be detailed for its protection. For instance, prior to 1970 the American mechanized battalion had only eight major ATW (defined as generally capable of defeating medium tanks at ranges of 1,000 metres or more) and 18 short-range *Bazooka*-type infantry ATW. The mechanized division inventory was 48 major and 117 infantry ATW (including 9 in the military police company). The armoured division had 40 major and 99 infantry ATW. Tanks, of which there were 351 in the armoured and 243 in the mechanized division, thus outnumbered all ATW by 2 to 1. (*Reference Book: The Division*, United States Army Command and Staff College, Fort Leavenworth,

The two critical aspects of this anti-tank deficiency are concentration of force and the effectiveness of the weapons. Technology only affects the latter aspect *directly*. Until recently anti-tank weapons had relatively poor performance except in special situations; this is reflected in the military cliché that the best anti-tank weapon is another tank. Technology has now provided four types of effective ATW.[48]

Technology cannot, however, solve the difficulty posed by concentration. Technology can simplify anti-tank weapons so that a greater number can be embedded into a combat unit; but only a limited number can be so absorbed without creating other weaknesses. Concentration calls for greater numbers of combat units and an increased ability to shift units from sector to sector. This is a doctrinal and organizational problem which technology can affect only *indirectly*, by easing the constraints that were instrumental in shaping these doctrines and organizations originally.

IV. DESIGNING FOR CONVENTIONAL EQUIVALENCE

In Europe today the tank is supreme. If it can be stopped, the ability to project force and to occupy territory will no longer exist. No other means of projecting force seems practical. Foot infantry is too slow-moving to be decisive. Paratroop infantry is too vulnerable, both en route and once landed, to be a major military threat other than in surprise attacks designed to sow confusion and to destroy high value targets like bridges and prepositioned stocks; its main usefulness is diversionary, even political, not warfighting. Helicopter infantry has most of the same limitations as paratroop infantry; though when the defender is in disarray, as after a surprise attack or an armoured breakthrough, helicopter infantry preceding the main attack force and using roadblock tactics can accentuate that disarray.

While an army is a complete organism, its core is the combined team built around the tank. In fact it can be said that the ultimate function of general-purpose forces in Europe, whether tactical air power, transport, sea-lane protection forces or the like, is to assist or immobilize the armoured vehicle – a fact easily forgotten as each component seeks to optimize its own contribution. The result is that not enough attention is paid to the key anti-tank function itself and too much is spent on some components which would have little relevance in the overall context if forces were organized differently.

Land warfare revolves around three sequential functions – attrition, holding and restoring. At Agincourt and Crecy, English long-bow men caused such casualties at long range to the French knights that the holding and restoring functions were minimal. In World War I the holding function was dominant. In World War II the tank and the tactical fighter broke the holding function and the restoring (or counter-attack) function became critical – as in

Kansas, April 1971.) By comparison a Soviet-type motorized rifle division had 36 major, 72 intermediate (i.e. crew-served anti-tank weapons of less than 90mm), 280 infantry ATW plus 215 tanks. The Soviet tank division had 6 major, 24 intermediate and 112 infantry ATW plus 343 tanks. (FM 30–102, Department of the Army, Washington, DC, October 1969). The Soviet Union has now upgraded her anti-tank weapons with anti-tank guided missiles (ATGM) and larger calibre guns. Thus, given that the Warsaw Pact had as many motorized divisions in central Europe as NATO had total divisions, the widely reported claim (and its implied significance) that NATO has a 50 per cent superiority in ATW over the Warsaw Pact could not possibly have been true (*How much is Enough, op. cit.*, p. 149). The cause of this erroneous weapons count was the use of 90mm as the cut-off calibre in counting anti-tank weapons. All anti-tank weapons 90mm or over were lumped together as equally effective; all below 90mm were excluded. This criterion counted American rifle-platoon weapons but excluded comparable but slightly lower calibre weapons of other nations; moreover it excluded many Soviet weapons in specialized anti-tank units. 90mm calibre is generally considered necessary for penetrating frontal tank armour, but, whereas American employment practice is based on this, Soviet forces, through their echeloned defences and their system of specialized anti-tank units, emphasize flanking fire, which has obvious advantages and can utilize smaller calibres.

The on-going European Defence Improvement Programme for upgrading anti-tank capabilities has helped to correct NATO's anti-tank deficiency and has given NATO a temporary advantage in this field. Numbers are now about equal for major anti-tank weapons and NATO's new family of ATGM is a generation more advanced than the Soviet inventory. But this qualitative superiority will fade when the Soviet Union introduces her own second generation ATGM, and the numbers gap will reappear as the Soviet Union mounts them on the guidance rails appearing on her latest armoured personnel carriers. With NATO's present structure, improvement programmes cannot permanently close this numbers gap because the Warsaw Pact has twice as many divisions and half again as many battalions per division.

[48] Immediately available are (1) second generation ATGM, (2) low recoil, flat trajectory, smooth-bore cannon, and (3) rapidly distributable minelets that can temporarily incapacitate vehicles. Available in the near future are precision guided ('smart') artillery and heavy mortars. A possibility for the further proliferation of cheap anti-tank weapons is the beam rider for smooth bore cannon, allowing cheap multi-purpose cannon to replace the ATGM for mechanized units. Offsetting the advance in anti-tank weapons and the notion of defeating an expensive tank with a cheap weapon is the possibility of low recoil, breechless, high velocity cannon which could enable tanks to be smaller and cheaper. Such tanks would emphasize fast-response direct fire, with beam riders and 'smart' artillery handling the longer ranges. These tanks do not need expensive ranging equipment to achieve greater accuracy for distances over 1,000 metres.

the tactic of the mobile defence. The ideal defence is to be able to destroy the opponent by attrition before he closes with the main holding or defending forces, but it is unlikely that this will ever be achieved, even with restructuring and new technology. What can be done is to redistribute the weight put into the three functions so that the defence can *hold* and is no longer dependent upon the function of last resort – *restoring*. Current technology used so as to remove constraints on the deployment of the defence can lead to significant savings and to the dominance of the defence over the offence. By the next decade sophisticated target acquisition and terminal guidance, albeit expensive, will be available to add an effective attrition capability at long range, thus helping to reduce casualties and further strengthen defence.[49]

Holding or stopping an opponent will probably always be the most important function of defence; even if the future brings effective attrition capability, forward defence is still necessary to gain time to prevent loss of territory and to deter even limited territorial seizures, which can be useful for political bargaining. Moreover, a counter-attack to restore lost territory can become less practicable because of the threat of nuclear escalation and because new technology can make *organized* defence increasingly difficult to overcome. In an alliance, forward defence is both a political and military necessity to protect and anchor its forward members.

The difficulty is how to organize a forward defence when NATO already has problems with a less demanding mobile defence concept – which is prepared to trade some forward space to give time and room for reserves to be deployed. A true forward defence has been considered militarily impracticable, since even a really strong wall can be penetrated so that the battle becomes one of manoeuvre in the rear areas.[50]

NATO's missing ingredient is its lack of large ready reserves on the battlefield; these are necessary for a viable defence in armoured warfare and for warfighting when tactical nuclear weapons might be used. With such reserves a defender can parry enemy penetration efforts and identify the critical ones, so that additional reserves can be brought up to the key sectors and adjacent sectors thinned to make further forces available for deepening the defence and for counter-attack. If a defending commander has enough reserves at his disposal, a mobile defence becomes a possibility.[51] But this is still not a forward defence, which requires large battlefield reserves and the removal of several technological constraints so that forward forces can be newly deployed into a chequerboard defence of numerous strong points *physically* occupying the forward area.

The following sections discuss (1) how these large reserves can be obtained from restructuring the present forces; (2) the way that new technological developments can make possible a change of deployment; and (3) the relative merits of the mobile and chequerboard defence concepts.

Restructuring for larger reserves

Restructuring to yield larger battlefield reserves requires major changes in tactics, organization, materiel and even peacetime personnel and training practices. But while this may seem a frightening array of problems to tackle, the concepts underlying such a restructuring are relatively straightforward and well-known. The difficulty is showing that a new meshing can yield a new, more effective force posture.

The pivotal issue is the nature of armoured warfare and the Soviet military doctrine. In armoured warfare major attacks are not made across the front in the manner of the Western Allies in World War II; instead the bulk of the front is held by deploying secondary, economy of force, units to deceive and to pin down opposing forces while the main attack is concentrated in one or more narrow sectors, to achieve a deep penetration and subsequent exploitation in the defence rear areas. This well-known doctrine has underlain some spectacular German, Soviet and Israeli military triumphs. What has been less recognized is that the defence can be symmetrically organized to counter these tactics. Just as the attacker should not dissipate his forces by distributing them across the front, neither should the defender. Just as the attacker concentrates his forces for intense combat in narrow sectors and

<hr>

[49] In addition to new anti-tank weapons, the constraint-removing technologies are (1) night vision and early-warning devices, and (2) cheap wheeled, instead of tracked, cross-country vehicles. Major economies can be derived from better integration of ground and air forces and from new developments in rapid-fire, long-range artillery. The new high-cost technology is largely associated with the elusive long-range attrition function; available in the next decade are such things as remotely piloted vehicles, sometimes called 'drones'; and various mid-course and terminal-guidance techniques for stand-off missiles like MTI radars and DME, as discussed later (see pp. 33–36).

[50] The standard defence concepts are linear defence (termed 'area' defence in the United States Army's lexicon) and mobile defence. In linear defence the bulk of the forces are placed forward in the line, and only small mobile reserves are retained. The mobile defence reverses this allocation of combat power: forward forces are thinned out and the brunt of the defence is borne by mobile reserves which are deployed when needed in a battle of manoeuvre. NATO has historically been deployed for a linear defence; in recent years, as the infantry's anti-tank capability has increased, the emphasis has shifted towards a mobile defence. A third concept,

discussed in detail in Part V, is the chequerboard defence. This defence derives its name from the game of chequers (draughts): the defence is widely distributed across the front and in depth in the form of small strongpoints.

[51] For a detailed explanation of the danger of trying to execute a mobile defence with inadequate reserves, see pp. 27–32.

holds strong forces in reserve, so too should the defender. Finally, just as the attacker can concentrate his logistic assets for the breakthrough effort, so too should the defender concentrate his logistic assets according to the varying demands of his forces in combat.

Anti-tank cavalry
The first step towards obtaining larger reserves requires the organizing of anti-tank cavalry units, relatively simply equipped, and substituting them for the expensive mechanized infantry and tank battalions now deployed in linear fashion across the front. Leaving such expensive assets in the line risks their being tied down by the attacker's deception efforts, whereas, with a strong anti-tank cavalry screen in front of them, these heavy battalions can be placed in the commander's local reserve. They can be thrown into the battle locally or in other sectors of the front as the situation demands.

A *limited* pull-back of forward mainline units is not a deviation from the politically necessary concept of forward defence. Since the first line of defence may be some distance from the border, the position to which units are moved can be located near present battle positions; the purpose of pulling-back combat units which are *presently* forward is to form *local* reserves. There are, of course, a number of areas on the central front which may be considered likely to attract enemy thrusts in strength, notably the high speed road approaches (perhaps three can be classified as major and another six as minor); in these sectors it will be necessary to deploy mainline battalions forward in a blocking position.[52] For the remainder of the front, anti-tank cavalry backed up by battalions in local reserve should be sufficient to hold an enemy himself concentrating on the main approaches.[53]

How to create an armoured cavalry with an anti-tank bite is therefore the first problem that technology must solve. NATO's present armoured cavalry and reconnaissance units are too few in numbers and are primarily designed to warn, not to fight; their anti-tank capability is weak. They should really be able to carry out the cavalry roles *and* double as anti-tank units.

The company-size organization that has existed in the United States Army since World War II has been composed of tank, scout, infantry and mortar elements, each of approximately platoon size per company. This organization is now dated. New technology in the form of anti-tank guided missiles (ATGM), low-pressure cannon[54] and wheeled cross-country traction now permits the tank and scout elements to be combined into a new anti-tank scout component.[55] By substituting long-range ATGM and intermediate-range low-pressure cannon for heavy high-velocity cannon, the heavy-chassis tank is no longer needed for cavalry functions. Similarly, since light-weight anti-tank systems are now available, scout vehicles can now mount a major anti-tank weapon, as well as retaining an anti-infantry capability. All vehicles can have light machine guns. The smooth-bore cannon is a dual-purpose, anti-infantry, anti-tank weapon; and vehicles mounting an ATGM can also mount a light cannon. With such an increase in anti-tank weapons and their growing importance, some emphasis can also be given to anti-tank suppression using multi-barrelled anti-aircraft guns, which are also effective against the thin-

---

[51] The enemy may well avoid attacking in the high speed corridors initially, but he must necessarily soon do so. A breakthrough elsewhere would be considerably less dangerous to NATO: slower movement would allow NATO more adjustment time and the terrain would be easier to defend.
[52] In American practice each corps sector is screened by an armoured cavalry regiment of three cavalry battalions and one tank battalion-equivalent. The problem is that the sectors are each too wide for the regiment's manpower and anti-tank strength. By the next decade, as equipment such as simplified and indirect stand-off anti-tank weapons are introduced, much of the screening and early blocking roles in anticipated secondary areas could be handled by militia forces. This would free a larger proportion of the better trained and heavier equipped active forces for the main battle. For a discussion of this possibility with militia forces see Kenneth Hunt, *The Alliance and Europe: Part II: Defence with Fewer Men*, Adelphi Paper, No. 98 (London: IISS, October 1973), pp. 37–40.

[54] Technically this is a high chamber-pressure, low muzzle-pressure, fin-stabilized gun system. The technical characteristics and value of the system are: its lightness; the ability to use ammunition not limited in its length and thus its penetrability by the length–calibre ratio demanded by the spin-stabilization of the high velocity gun; low recoil; and relatively flat trajectory (that in the case of the French 90mm and a moderate ranging capability can yield good first hit probabilities up to 1,000 metres against tanks). This is a critical range against Soviet-style attacks. For example, Soviet tank gunnery with its very high muzzle velocities is designed for fast first-hit probabilities for ranges up to 1,000 metres. Thereafter accuracy falls off rapidly (i.e. from about 0·8 probability at 800 metres to only about 0·5 at 1,000 metres) because of rudimentary fire control. (In World War II the mean tank-battle distance was about 800 metres.) The new Russian APC also has a 73mm low-pressure cannon but its short tube-length and relatively small diameter plus the fact that there is also the ability to launch ATGM suggests it has been designed for close-in infantry support against moderately hard targets and for anti-tank suppression.
[55] The United States Army has recently announced the opposite approach. Apparently recognizing the anti-tank weakness of its armoured cavalry, the United States Army in Europe began in January 1973 to convert the armoured cavalry's scout section into an additional light tank section. This essentially converts the armoured cavalry into a light tank unit and forgoes much of its still required scouting role. Moreover, the light tank being used – the *Sheridan* – is an expensive, plague-ridden weapons system that exemplifies how technology has been misused in trying to satisfy questionable specifications (specifically a stress on aluminium for lightness and trying to combine a gun and a missile in a single launcher for simplicity).

skinned armoured segment of a tank force and in close-in fighting, such as at night and in woods and built-up urban areas. Finally, wheeled technology as developed by France and Britain has reached the point where armoured cars approach the cross-country mobility of tracked vehicles on European terrain.

Replacing heavy-chassis tanks with large numbers of cheaper anti-tank armoured cars has attractions in combining the anti-tank and reconnaissance functions in one unit. If anti-tank cavalry is also to substitute for mainline battalions in holding terrain in secondary sectors it needs organic infantry, but to be economical must keep its establishment streamlined. While British and French cavalry units have had no organic infantry, the American organization of three regular infantry squads per cavalry company is too lavish. Two reduced-size squads will suffice, and they can also handle sapper duties.[56] Finally the American company's three heavy mortars are redundant. Close-in defence can be provided by grenade and smoke dispensers on the vehicles, as in British units, and indirect firepower can be provided by supporting artillery.

Substituting a quieter and smaller wheeled vehicle offers new tactical possibilities, particularly that of restoring to cavalry some of its traditional harassing and ambushing roles against tanks as well as against infantry. With the small addition of sapper infantry, cavalry can add more variety to its roles and act as an economical substitute for mainline battalions in secondary sectors. For the screening role a cavalry company of 90 men with as many as 24 major anti-tank weapons could readily substitute for a mainline battalion.[57] If the enemy does attack in strength, the cavalry has enough anti-tank and anti-infantry strength to block temporarily while a mainline battalion held in local reserve deploys. Thereafter the anti-tank cavalry can reinforce other cavalry on the flanks of the main battle or concentrate as an anti-tank blocking force behind the main battalions.

The combination of the above changes means that in the case of the American armoured cavalry, for example, the number of major anti-tank weapons could be tripled, while equipment and maintenance costs could be halved, manpower reduced by a third, and operational flexibility increased. This exemplifies a theme of this Paper: *Simple technology accompanied by doctrinal and organizational changes can often lead to significant increases in cost effectiveness.*

Concentrating logistics

Step 2 in producing larger battlefield reserves involves reducing the size of the divisional slice by as much as half, and using the manpower thus saved to form cadre divisions. Pulling back the bulk of the mainline battalions from the front line to a reserve position also serves this purpose since it implies reduced combat activity and therefore lower logistic demands.

The way to cut the divisional slice is essentially by greater pooling of logistic resources, so that they can be used in a support framework to serve the formations that will be heavily committed – the Soviet practice – rather than dissipating them almost equally across the front. Since much of the combat, in the less important sectors, is likely to be in the nature of holding actions, logistic demands will be lighter and thus logistic elements can be fewer; logistic resources can be concentrated around those parts of the front where enemy thrusts are centred.

These resources need to be available to a division in strength only when it is actively committed. Rather than assigning logistic assets organically (i.e. as integral parts of a formal organization) all the way down to company, battalion and division level on the basis of what is required for sustained combat, they need to be taken out and organizations streamlined to no more than is required to support the unit at a low activity level. The assets withdrawn can then be concentrated under higher headquarters for supporting formations when actively engaged. Since only part of the force will be engaged heavily at any one time and no unit will be engaged on a sustained basis, the stripped logistic assets will exceed logistic requirements, enabling much of the released resources to be used for other purposes, notably for forming additional, cadre, divisions. By altering logistic practices (and, of course, the operating and tactical practices that dictated them), NATO's wartime divisional slice can be roughly halved in size, while giving the same 'foxhole' combat strength.[58]

[56] A sapper is the British term for a light field engineer, particularly useful in countering a Soviet-style attack. Soviet forces use roads to obtain speed and to reduce logistic demands. Except for the manoeuvring of their most forward units, tank units prefer roads, so as to reduce (1) fuel consumption (fuel consumption off roads is 2·5 times higher); (2) maintenance (off-road use greatly increases major overhauling requirements and the 'rugged' Soviet tank in any case requires major overhauling sooner than its Western counterpart); and (3) crew fatigue. Soviet logistic support trucks are usually of a non-tactical nature with very limited off-road capabilities (as is their artillery). Accordingly a series of obstacles along each road and track, with an occasional supporting minefield, form a cheap way of disrupting Soviet momentum.

[57] The standard H-series tables of organization and equipment for an American armoured cavalry company is 161 men with 9 light tanks and 15 20mm armed reconnaissance tracked vehicles. An American mechanized infantry battalion has 888 men and 18 major ATGM.

For a description of possible tables of organization and equipment for anti-tank cavalry, defensive infantry and strike-oriented tank battalions, see Canby, *op. cit.* in note 40.

[58] Such ubiquitous logistic services as engineer, maintenance, transport, medical and food services account for 40 per cent of the personnel in an American-style theatre force. Signal

This requires fundamental changes in the principles underlying NATO army organization. Foremost it means giving up the concept of *logistic balance*; but a field army ought not to be organized on the basis that all units can be fought, sustained and supported simultaneously. Combat units would be sustained by 'support frameworks', into which they would be 'plugged', like electric motors into a wall socket. This corresponds with differences in casualty rates, which, as has been said above, are far higher for men in infantry and tank platoons than for logistic troops. Placing logistic resources in every line unit to support sustained combat is like matching speartips and spearshafts on a one-to-one basis: when the speartip is lost, the shaft is rendered useless. Support frameworks permit the maintenance of several speartips per shaft, even though they cannot all be used simultaneously. The criterion is not the ability to sustain every unit all of the time, but to support the total force as required.

Units with extensive organic support tend towards an expeditionary force posture that relies upon its own resources rather than upon any local assets that may be available. Decentralization and distribution of logistic assets to commanders preoccupied with an urgent combat task inevitably leads to a slighting of logistics and an attitude hostile to the uncertainties inherent in mobilizing local resources. If, however, logistics are centralized under specialized logistic commanders whose atten-

tion is focused primarily on logistics, openings may be created whereby commercial or other assets can be used to advantage (as at the moment, for example, with fuel pipelines). In the NATO area there are many resources within the civilian economy that might be tapped in wartime to supplement and flesh out active forces: transport, medical, engineer construction and signal communications spring immediately to mind.

A new replacement system

A third step that might be taken, and a logical follow-on to logistic concentration, would be to change the replacement system from an individual to a predominantly unit-based one. At present NATO combat units would replace their battle casualties by individuals sent to join them, and the unit size is governed by the fact that it must be strong enough to absorb casualties and still function while waiting for replacements. Units designed to be replaced complete must be withdrawn once they sustain casualties beyond a moderate level; their size or staying power is not such as to enable them to continue fighting.

Individual replacement assumes that personnel (and equipment) can be fed into units as losses slowly build up. This is unlikely to be feasible in high intensity combat; at the extreme, tactical nuclear war would rule this out completely, because the large numbers involved could not be effectively

communications and clerical-type services account for another 25 per cent. For ways in which each of these services can be re-organized for outright reductions and through meshing with mobilized local resources in wartime, see S. Canby in R. Komer (ed.) *Restructuring NATO Forces to Compensate for MBFR* (Santa Monica: Rand Corporation, Nov. 1973), and S. Canby and R. Rainey, *Restructuring of US NATO Ground Forces: The Division* (Rand Corporation, Oct. 1972).

Two examples illustrate these re-organizations: that of engineering and maintenance which form 10 and 15 per cent of the force respectively. Engineering services are focused upon rearward construction, bridging, line of communication maintenance and barrier construction/breaching. In NATO's present circumstances of marked inferiority, most of these services have little value – NATO's task would be destruction so as to assist a withdrawal, not constructing or maintaining a transport network. Elaborate barriers are demanding in time and materials and would have little value; barriers must be physically defended to be effective, and once penetrated become outflanked. (Protective obstacles for local defence remain important, but these can be readily handled by the new family of artillery-delivered minelets.) In such circumstances NATO's basic engineer requirement is for simple sapper-type engineers, designed to disrupt in depth the European transport grid by repeated demolitions and short flanking minefields. The bridging requirement – to ensure friendly troops are not trapped – can be readily handled by armoured vehicle-launched bridges (AVLB); in the few cases where these do not have the required reach, extra bridging equipment can be prepositioned or abutments made in peacetime to support multiple AVLB.

If NATO can stabilize the battle, then elaborate barriers are again unnecessary, and the engineer function behind the

forward divisions can be handled by mobilized civil assets, a function the German Territorial Force could fulfil. Sapper engineers are again the most suitable for the division, both in the defence and for the breaching of obstacles in the offence. Only a relatively few more generalized combat engineers are required, and these can be pooled, largely for use in the penetration sectors. In short, shedding the expeditionary force concept allows NATO, and the US Army in particular, a 60 per cent reduction in present engineer strength and an even larger one in the fleshed-out wartime slice.

Whereas engineer strengths could be reduced by substituting civil assets and challenging the tasks performed, about a third of total maintenance personnel are organic to non-maintenance units, and maintenance cuts in general cannot be made without significantly affecting the remainder of the force. Hence maintenance reductions must either reduce force activity, or the force must revise its operating practices. The latter can be done without reducing (1) quality of training, (2) level of equipment readiness, and (3) combat sustainability, by (a) lowering the *demand* for maintenance services by revising personnel replacement and training practices and by garaging equipment, and (b) increasing the effective *supply* of maintenance services by emphasizing direct support contact teams, substituting civilians for military personnel, and consolidating all direct support, general support and depot maintenance (and corresponding supply responsibility) under a specialized, high-level command in peacetime. These changes can reduce present peacetime military maintenance strengths by 55 per cent. This reduced maintenance strength would have to be augmented by 30 per cent in wartime, with the increase concentrated in direct support maintenance units for allocation by higher headquarters to the heavy combat penetration battles.

absorbed. In intense conventional fighting heavily hit units would have to be pulled out for regrouping and cohesive fresh units be brought in to replace them. Moreover, if losses are high, the combat portion of units is likely to be destroyed while their logistic component survives.

Of the two systems, individual replacement is more attractive in protracted infantry wars, where combat is diffused among all line units. Thus, individual replacement systems are most appropriate under conditions such as those of World War II, Korea and Vietnam, where sustained conflict and a tactical philosophy allocate effort (and supplies) more or less equally across the front. If losses are low enough so that replacements can be readily absorbed, maintaining an experienced and functioning team is a realistic and desirable objective. The key, of course, is the assumption that there will be few or only gradual casualties.

Unit replacement seems the more suitable system for tactical nuclear conflict, short conventional wars and the focused penetration/exploitation tactics of *blitzkrieg*-style fighting.

In making such a change, as with stripping out much organic logistics, the emphasis is shifted from firepower, staying power and sustained strength to shock-power and 'surgeability'.[59] Staying power (an ability to absorb casualties and still function) and unit sustainability lead to large units. The larger the unit, the greater become the problems of control, road congestion and movement.[60] While a doctrine

[59] To the author's knowledge, the distinction between firepower and shock-power has never been carefully defined. One distinction is the relative emphasis on fire and manoeuvre. According to this distinction, a firepower-style army relies extensively on indirect fire support (i.e. mortars, artillery and air delivery) to destroy or neutralize the enemy by fire and to allow friendly forces to close. A shock-power army emphasizes manoeuvre to defeat the enemy and a higher proportion of *direct* firepower (tanks, assault guns, etc.) accompanying manoeuvre units. A second and better distinction, in the author's opinion, is between brute force and psychological impact. A firepower army tries to overwhelm its opponent by the sheer weight of its resources, as on the Western Front in World War I and in 1944-45, more recently in Vietnam, and in the quantitative models in vogue in the analytical community. A shock-power army tries to create the impression of great strength, often by reliance on mobility to augment its force by speed and momentum (and deception) so as to sap the morale of its opponent and to paralyse his command and control. Two classic examples of shock-power armies are Genghis Khan's Golden Horde and the German World War II *blitzkrieg*.

'Surgeability' is a stress upon initial hitting power that cannot be sustained, like an emphasis upon automatic weapons – a feature of Communist armies that cost the United States heavily in Korea and Vietnam. Surgeability is normally inversely related to staying power. Surgeability units are designed for maximum short-run impact; they are therefore designed to be small and to be fought with the expectation that casualty rates will be high but that combat will be sporadic or that the unit will soon be replaced.

[60] General Ralph Haines, when Commanding General of

which calls for linear deployment inherently stresses firepower (and demanding logistic lines of communication) because of need to support strung-out forces, size *per se* also inhibits movement and leads to substituting firepower for manoeuvre. Replacement units, on the other hand, because of their smaller size and more limited period of combat at any one time, tend to emphasize surgeability, manoeuvrability and shock-power.

Forming cadre divisions

Centralizing logistics and replacement by units logically leads to using manpower savings and reserve manpower to form cadre units rather than manpower pools, thus enabling a larger number of divisions to be fielded. Large equipment and war reserve stocks and manpower pools, standing by until losses create a demand, have little utility in a short war, particularly if the attacker organizes his reserves to provide large numbers of replacement divisions. Such a situation creates an asymmetry. A structure designed to sustain units may well be better in a long war, but in a short war the surging forces may win by swamping the defender's smaller structure. The forces of the European NATO countries on the continent must therefore be organized to benefit from rapid mobilization; the forces of the Allies outside continental Europe must be designed for rapid absorption of reinforcements while still retaining the ability to deploy before these reinforcements arrive.

If NATO countries concentrate their logistics and take the other steps that have been described above, manpower savings of about half could be made in *wartime* slices.[61] If the peacetime strength assigned to this redundant structure-is then used to create full strength and cadre divisions, NATO's present divisional structure could well be tripled. For instance, by going to three categories of readiness as the Soviet Union does,[62] the four continental countries (France, Germany, Belgium and Holland) could go from their present 22 division structure to a 60 division structure within the constraints of present active and reserve strengths (15 Category I, 20 Category II and 25 Category III divisions; other combinations are of course possible).[63] The chief constraint is having

United States Continental Army Command, vividly described this phenomenon as follows: 'We have become so vast (at brigade and division levels) we may trip over our own entrails, while trying to manoeuvre on the battlefield.' 'Haines Warns of Tripping on Trivia', *Army Times*, 7 June 1972.

[61] See S. Canby, *op. cit.* in note 58, and S. Canby and R. Rainey, *op. cit.* in note 58. The Soviet Union, of course, already has divisional slices – as adjusted for equal foxhole strength – of this size. Soviet slices moreover are heavy in tanks and artillery, whereas NATO lays emphasis on infantry and mortars.

[62] See note 38.

[63] For instance, German ground forces total 340,000 in peacetime, of which 35,000 are in the territorial forces and

23

enough Category I divisions to meet the case of the surprise attack, to block an enemy advance and gain sufficient time – several days – for the Category II divisions to mobilize and to deploy. Fifteen such Category I divisions, added to the likely American, British and Canadian formations, which need to be maintained in Category I for rapid absorption of reinforcements, would give NATO as many high readiness Category I divisions as at present; 20 Category II divisions would give NATO a large reserve within days (particularly if divisional tasks were simplified as discussed in the next section).

Similarly, the United States Army in Europe – and to a lesser extent the BAOR – could triple its present combatant strength in a 14-day period. The restructuring measures outlined above would permit the present United States Army strength in Germany to be transformed from 4½ divisions into 9⅔ division equivalents.[64] If the forces were in addition organized for rapid deployment, a 14-division-equivalent structure could be maintained in the theatre.[65]

American reinforcements and forward stocks

Despite the considerable airlift capacity of the United States, there are problems that impose limits on rapid reinforcement. These are largely caused by bottlenecks inherent in the plan to reinforce with large units: difficulties over keeping formations ready in the United States, absorbing them on arrival in Europe and giving them time to shake

down and make themselves ready for operations after that.

A concept of reinforcing by components or 'packages' of 100–200 men each could eliminate these bottlenecks.[66] Since there would be a large number involved and readiness could be rotated, some units could always be at high readiness in the United States, thus both avoiding delay and reducing hardship in peacetime. If their equipment was pre-positioned with a parent formation, instead of in a supply depot in Kaiserslautern, the bottleneck at the depot could be avoided and air terminals near the parent unit could be used. Since the arrivals would be marrying with a functioning parent unit, they would take less time to be ready for combat.

Organizationally this procedure would involve increasing the number of companies per battalion from three to four (and adding two battalions per division); all equipment would be prepositioned with the parent battalion in Europe, but the fourth company plus part of the battalion overheads would be assigned to a division based in the United States. The European-based division's assistant commander (a general officer) and the corresponding deputy commander at brigade and battalion level would command the sub-units designed to reinforce the parent unit in war, but located in peacetime in the United States. Reserve readiness would also be enhanced by assigning corresponding units from the reserve forces to replace the active force packages when they depart in wartime, thus keeping the United States-based division up to strength.

Generally the concept is that, in peacetime, infantry and tank battalions would continue to deploy three line-companies in Europe. While each battalion would be at lower than full wartime strength, it would have nearly as much combat strength and considerably more anti-tank capability than such units now have in Europe.[67] Its wartime effectiveness would, of course, be enhanced when the fourth company and the remaining battalion elements arrived. In addition to the extra combat power embodied in the fourth company, flexibility would be increased, since combinations of four companies permit a much wider range of tactical options while spreading overheads over more units. Flexibility and the additional company are particularly important if the combat-force density is as low as in Europe. With available communications, battalions should be fully capable of controlling the additional company or indeed companies; armoured cavalry battalions, which have a greater command and control problem than tank and infantry battalions, already have four manoeuvre companies. Smaller battalions with more line-companies would also be

28,000 in the various defence agencies. Assuming wartime division slices of 20,000, the German field army of 277,000 could produce a 25 division structure, say 7 Category I division slices at near full strength; 8 Category II at 50 per cent strength; and 10 Category III at 25 per cent strength. With a similar cadre ratio and assigned strength in the defence agencies, France could support 23 divisions, the Netherlands 6½ and Belgium 5½. Another combination to increase immediate front-line readiness and overall structure, while reducing German visibility, would be to lower the German contribution to 17 higher readiness divisions (10 Category I and 7 Category II), and lower the French and Benelux cadre ratio, increasing their division structure from the 35 above to 51 divisions, to correspond with their more rearward geographical position in peacetime. Personnel in the defence agencies can also be used to enrich the active force content of the cadre units upon mobilization as is done in Israel.
[64] Technically this would be 8 division flags (exclusive of 2 armoured cavalry regiments) plus 21 additional manoeuvre battalions (there are 12 in a division).
[65] Technically this would be 12 division flags, with a peacetime combat strength of 9 division equivalents. The wartime strength would however equate to 14 division equivalents. At the moment there are only 11 sets of division-equivalent equipment in Europe: 4½ with the stationed divisions; 2⅔ sets prepositioned for reinforcing formations; about 2 sets in war reserve stocks, and a further 2 sets which could be obtained from streamlining and consolidating within divisions. For details see S. Canby *op. cit.* in note 58. Similar procedures would yield a 6½ division British/Canadian structure.
[66] See S. Canby and R. Rainey, *op. cit.* in note 58, pp. 10–12.
[67] *Ibid.*, pp. 20–47.

advantageous in tactical nuclear operations because of the greater dispersion afforded and their smaller target size and detection signature.

Forward storage of equipment with four-company battalions would, furthermore, reduce in-theatre deployment times, decrease the vulnerability of the prepositioned equipment stocks along the Mannheim–Kaiserslautern *autobahn* and, together with changes in training practices, permit reductions in maintenance personnel. The present arrangements cause bottlenecks: divisions flown in at full strength must be funnelled through the equipment depots and, after marrying personnel and equipment, need time to adjust to the new environment and deploy forward. The larger the unit, the greater the readjustment time. Battalion packages would reduce these problems. The men could be flown by chartered commercial jets to a number of civilian airfields throughout Southern Germany and bussed the relatively short distances to their units – thus substantially increasing USAREUR's absorption capacity. Since the units are company-size, confusion could thus be minimized and reinforcements rapidly deployed forward.

The major argument against this concept is the alleged vulnerability of equipment stocks placed forward with a parent unit. The other forms of vulnerability, however, are unambiguously eliminated – the vulnerability of the large storage sites to air attack or surprise airborne assault. Forward storage is said to be vulnerable because NATO is unlikely to be able to hold forward. However, restructuring, of which this sub-unit deployment plan for American, British and Canadian forces is but one part, is designed to correct this deficiency so that NATO *can* hold forward. In the American sector alone the peacetime equivalent force, at present theatre strengths, would be 9 division equivalents – double the present number of combat troops. If this force were unable to hold, along with allies similarly strengthened, for the two week period required to complete reinforcement, then NATO's present posture would certainly be inadequate and Warsaw Pact forces could reach the Rhine in days, cutting off large NATO forces. The Rhine would then be an inadequately defended obstacle and stockpiles behind it would be lost before any reinforcements from the United States could arrive. What happens thereafter is somewhat moot, for *the ground war would have been effectively lost.*

Summary
This section has indicated how NATO could almost triple its number of division equivalents from 29 to 80[68] with the same numbers of men, active and reserve, as now. This would be produced by centralization of logistics (with acceptance of logistic imbalance); replacement by units rather than individuals; increased reliance on civilian logistic assets; a cadre system to form new divisions; and reinforcement by battalion packages rather than by larger formations for American, British and Canadian forces. Shock-power would take precedence over firepower. Though no more men would be needed, there would be a one-time requirement to provide equipment for the additional divisions raised: perhaps 35 sets costing some \$5–\$7·5 billion.[69]

Wartime divisional slices would be drastically reduced to about half, comparable with those of the Soviet Union. Just as the Soviet Union has 20 divisions in the Group of Soviet Forces, Germany, in a high state of readiness, so NATO can maintain a similar number of ready divisions and back them with a large number in reserve.

Restructuring could therefore give NATO the combat forces to match the Warsaw Pact conventionally. The next section is devoted to showing how new technology enables such forces to be deployed in new ways to give the defence advantage over the offence.

---

[68] Since Western divisions are about 25 per cent larger than their Soviet counterparts, this equates to about 100 Warsaw Pact divisions.

[69] The calculation is as follows: (1) NATO presently has on the central front (including prepositioned war reserves and assuming some streamlining) around 45 division sets of equipment, leaving a shortfall of 35 sets. (2) In general, equipment for the non-combat battalions already exists. (3) As will be discussed later, equipment can be simplified by adopting specialized counter-attack and ground-holding divisions. NATO already has the tanks and armoured personnel carriers for the more demanding counter-attack divisions. (4) Equipment costs for an American mechanized division of 6 mechanized and 5 tank battalions is \$240 million for the whole division and \$130 million for the 11 manoeuvre battalions (see *Army Force Planning Cost Handbook*, Comptroller of the Army, July 1973). (5) With the above assumptions, estimates range from \$4·6 billion to \$8·4 billion. Equipment simplifications suggest a cost of around \$4½ billion. (6) For the 45 existing division sets, equipment modifications of \$0·75 billion seem in order. A large purchase of ammunition is also required, but excluded here. Its size is sensitive to assumptions arising from the October War in the Middle East and the trade-off between conventional and precision-guided ordnance.

V. DEPLOYING FOR DEFENSIVE SUPERIORITY

Small strongpoints

If NATO generated more divisions and thus more operational reserves, it would gain flexibility, could opt for a mobile defence and match the Warsaw Pact conventionally. But this would only gain symmetry, not defensive dominance: dominance requires defence in depth against enemy penetration attempts. Even with restructuring and a cadre-mobilization system, a defence in depth provided by physical occupation of ground would still be infeasible because there would not be enough combat forces. Even if they did exist they would be vulnerable to nuclear weapons, *unless* based on much smaller combat groupings than now: strongpoints of company size – not battalion size as at present. With companies as viable defensive entities, a chequerboard defence (see below), which will give strength in depth, is possible. To achieve it requires the correct use of current technology and tactical and organizational change to match this.

The problem to be solved is how to enable isolated companies to hold against tanks and surprise infantry attacks in conditions of poor visibility. A brief review of United States Army reorganizations illustrates the point. The World War II organization required strongpoints (when used) of reinforced battalion size – almost a thousand men. In 1956 the Pentomic division used battle group strongpoints of almost 2,000 men. In 1962 the Reorganized Objective Army Division reverted to the battalion organization now mechanized (in Europe) via the armoured personnel carrier (APC) and having much increased firepower with the APC armament, a new small-arms family, heavier artillery calibres (155mm instead of 105mm) and artillery cluster bomblets. Even with such a large increase in firepower and mobility it was still not possible to rely on company-size strongpoints, and the defensive doctrine, except for purely delaying operations, remained that of companies coiled into a battalion perimeter or mutually supporting each other on a defensive line. Moreover, since a division slice of 40,000 only contained 10–11 line battalions, of which some had to be held in reserve, only 6 to 8 battalions were available for these strongpoints. NATO obviously did not have enough battalions for this type of defence; the loss of only a few strongpoints would have risked an enemy armoured breakthrough.

The missing ingredients were sufficient anti-tank and close-in night defences, but inexpensive current technology can now provide these and make it possible to have company- or even half-company-size strongpoints. Until now the only really effective anti-tank weapon has been the high-velocity gun

mounted on a protected mobile chassis. The inadequacy of anti-tank weapons as such has required that a tank company support each infantry battalion, and has led to tank-infantry task forces in forward defence as well as in counter-attack reserves. First generation ATGM were inaccurate, vulnerable and unreliable. Recoilless rifles (RCL) had poor first-hit probabilities, a tell-tale backblast and no protection for their crews. Light weapons like the 3·5 in. rocket launcher were inaccurate, awkward, short-range and in short supply (a foot-infantry heritage).

On-the-shelf technology · can change all this; ATGM such as the American *TOW*, with effective ranges double those of tank guns, have finally become available. RCL can now be replaced by high-/low-pressure cannons with fin-stabilized, relatively flat trajectories that give good first-hit probabilities up to 1,000 metres. Because of their lightness and low recoil, both weapons can be mounted on light armoured vehicles, possibly wheeled, at least in Europe where terrain conditions are not demanding. Good short-range, compact, infantry anti-tank weapons, like the German *Armbrust*, are also now available: one could literally be placed in every foxhole as an auxiliary weapon and stored in the APC when not needed. Finally, minelets and laser/infrared guidance can make artillery and heavy mortars a major tank immobilizer and killer.

Night-vision image-intensifiers and close-in early warning devices and sensors, highly developed in Vietnam, are the remaining ingredients that will enable small company units to become viable defensive entities, able to protect themselves in poor visibility and even see far enough to fire anti-tank weapons at night.

Thus with new technology, large battalion-size strongpoints are no longer necessary. If ATGM and low-pressure cannons were mounted on APC and armoured cars, a 650-man armoured infantry battalion of one anti-tank cavalry and four infantry companies could mount as many as 70 major anti-tank weapons, and be supported additionally by the new tank-killing capacity of artillery and mortars.[70]

[70] For organizational details, see S. Canby, *op. cit.* in note 57, pp. 10–12. The present American battalion has 3 infantry companies, an anti-tank and a reconnaissance platoon; its full strength is 888 men. Before 1972 it had a total of 8 major anti-tank weapons and 18 infantry anti-tank weapons. It now has 18 *TOW* ATGM and will soon replace the 18 infantry 90mm RCL with 27 intermediate range *Dragon* ATGM on the basis of one per squad. This will give the battalion 45 major anti-tank weapons of significantly improved quality.

Nevertheless a number of arguments can be marshalled against such dependence upon a single weapon type. Foremost, the ATGM is not designed to cope with the Soviet tactic of rushing the defence, using extensive suppressing fire to

Minor infantry anti-tank weapons in rifle squads could similarly be increased from their present two per platoon or 18 per (American) mechanized battalion to at least one per squad for a minimum of 38 for the revised battalion. Such an increase in anti-tank firepower is even more striking when it is recalled that the pre-1970 major anti-tank systems were relatively ineffective jeep-mounted 106mm RCL and *ENTAC* ATGM. In short, with night-vision devices and with companies having a greater anti-tank capacity than a whole mechanized battalion before 1970, a battalion can now be broken into *at least* four dispersed company strongpoints.

There are five significant implications of this change to a strongpoint defence. First, expensive tanks are no longer needed to protect infantry, leaving more for the counter-attack force. Second, armoured infantry can be split into two types: defensive infantry needing only wheeled APC, and attack infantry designed to accompany and support tanks – thus the only type requiring expensive tracked equipment.[71] Third, the roles of infantry and armoured cavalry are being shifted from an anti-infantry to an anti-vehicle orientation and can be given, as a spillover, an enhanced anti-aircraft capability. While the Warsaw Pact does not have a Western-style air force and its close air-support potential should not be exaggerated, the new-found ability to proliferate and diversify effective anti-tank weapons implies a need for weapons that can be used to suppress enemy anti-tank weapons. For this, a weapon like the Soviet four-barrelled 23mm anti-aircraft cannon is ideal. Trading off some of the anti-tank weapons for anti-aircraft guns does not detract from the anti-tank role, because Soviet tank attacks

will include both tanks and vehicles only lightly armoured. For defensive infantry, multiple automatic cannons are invaluable not only for air defence but also for night combat and defence against light armoured fighting vehicles (AFV). For the counter-attack forces, these weapons provide air cover during movement, when a ground unit is most vulnerable to air attack, firepower to suppress the anti-tank guided missile and a more effective weapon than the large tank gun for close-in fighting against infantry. Fourth, simplifying the mission and equipment of the bulk of the combat units also simplifies training and so facilitates a greater emphasis upon mobilization of reserves; it also eases the problems produced by short conscription periods and the possibly lower quality of recruits in an all-volunteer force. Fifth, if NATO should restructure itself along these lines, it already has enough tanks and tracked APC in its active and war reserve stocks; the additional armoured vehicles needed for increasing the number of combat units are only relatively inexpensive wheeled ones.

The chequerboard concept: defence by small strongpoints

For tactical nuclear warfare, a chequerboard-type defence combining dispersion and depth has long been recognized as offering the best defence. Dispersion is required to decrease the effects of nuclear weapons. Depth is necessary because any single line or combination of lines can be easily penetrated (particularly using nuclear weapons) and armour passed through; only a defence physically occupied in depth can bog down an armoured penetration force until it is weakened and slowed sufficiently for a counter-attack to destroy and eject it. Yet, though a chequerboard defence is needed for tactical nuclear warfare and has seemed desirable for conventional armoured warfare, it has not been adopted because of its manpower demands. Indeed a chequerboard defence across the breadth of West Germany is not feasible. But, as has been argued earlier, an elaborate defence need not be deployed everywhere on the central front, and NATO's combat divisions can also be practically tripled in number. The combination of these two factors makes the chequerboard concept no longer merely a desirable concept

cover and to mask the exposed leg of the dash. Coping with this tactic requires a high rate of fire, cover for the crew (the new European ATGM have this feature but the new American ones do not) and virtually no dead space in front of the weapon (the minimum range of ATGM is such that units depending on them have no anti-tank defence in very close-in fighting). A second group of deficiencies is the relative ease of combating the ATGM. Artillery can kill the crews of unprotected or unarmoured ATGM and, even if the crew is protected, cause it to lose control of the missile in flight. The slow time of flight of the missile also gives tanks time to take evasive action and to return fire, and automatic control, which makes the second generation system ATGM so attractive, can be spoofed. A third group of deficiencies is their limited usefulness against infantry targets in city fighting and in wooded areas (half of Germany is forested and much of the remainder is being overtaken by urban sprawl).

The major advantages of the ATGM are its lightness and long range, though the importance of range can be easily exaggerated in European weather and terrain conditions. Against this must be matched high missile cost and therefore limited confidence-building training for the crew, and the range potential of 'smart' artillery and beam-riding cannon fire.

[71] Such specialization of the infantry into an attack and defensive infantry is analogous to the German system itself

in World War II. Whereas 26 per cent of the United States Army order of battle in Europe consisted of armoured divisions, allocated on the basis of one per infantry corps, panzer and panzer grenadier divisions comprised only 18 per cent of the German order of battle. In addition, the United States Army assigned large numbers of tanks directly to infantry divisions, but Germany did not. (T. D. Stamps and V. J. Esposito, *A Military History of World War II*, vol. I, New York: United States Military Academy, West Point, 1953, and *The US Army in World War II: The Organization of Ground Combat Troops*, Washington, DC: Historical Division, Department of the Army, 1947.)

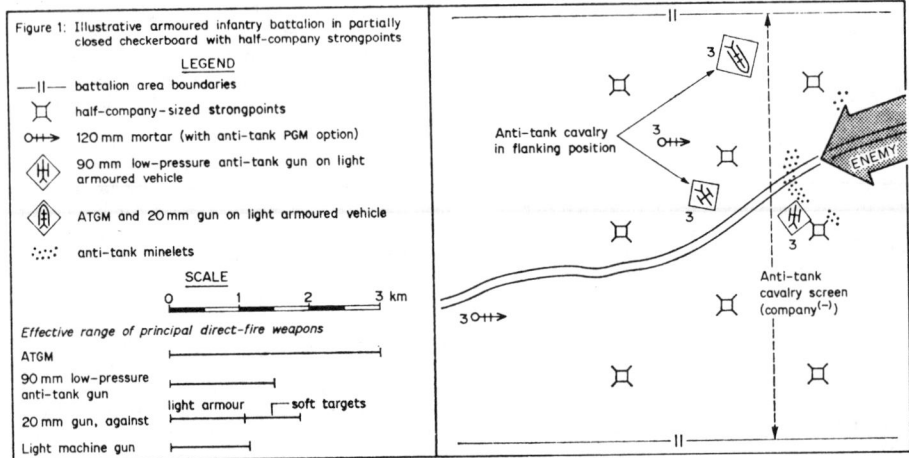

Figure 1: Illustrative armoured infantry battalion in partially closed checkerboard with half-company strongpoints

LEGEND

—‖— battalion area boundaries

half-company-sized strongpoints

120 mm mortar (with anti-tank PGM option)

90 mm low-pressure anti-tank gun on light armoured vehicle

ATGM and 20 mm gun on light armoured vehicle

anti-tank minelets

SCALE

0   1   2   3 km

Effective range of principal direct-fire weapons

ATGM

90 mm low-pressure anti-tank gun

20 mm gun, against   light armour   soft targets

Light machine gun

Anti-tank cavalry in flanking position

ENEMY

Anti-tank cavalry screen (company(-))

but a practicable one, as the following examples illustrate:

(1) Under NATO's present organization a 40,000-man division slice could provide about 10 battalion strongpoints. With restructuring, and removing the constraints upon small unit viability, 40,000 men can be shaped into two division slices with each slice organizing company-size strongpoints for a total of 80 strongpoints.

(2) With individual replacement to sustain an essentially fixed-size peacetime force, NATO would have difficulty in replacing units lost or decimated by Soviet swamping tactics or by tactical nuclear weapons. With unit replacement, it is less difficult to replace strongpoints. By creating extra divisions as has been advocated above, strongpoints could be increased further by an additional 40 per cent, making perhaps 110 for each 10 now available. Indeed, if part of the mobilization system were explicitly designed for strongpoint replacement, the ratio of 11 to 1 could be increased.

(3) Restructuring can give NATO 80 divisions. If 60 were held in reserve and 15 were placed near the border for forward defence, in addition to a strong 5-division anti-tank cavalry screen, the average front-age (exclusive of cavalry) would be 50km per forward division. Because of differences in threat and terrain, divisional sectors like those in Bavaria might initially be 80km wide, while sectors between Fulda and Hanover might be as low as 30km.[72] Assuming

good defensive terrain in secondary sectors, as is probable in Bavaria, a defensive division could physically chequerboard a sector 50km wide and 8km deep, as in Figure 1. Under these optimum conditions of (a) obstacles across the front, (b) good observation and fields of fire, and (c) low threat, half-company strongpoints averaging 2·5km diagonally apart and screened with anti-tank cavalry should be sufficient for containing enemy movement. In the more dangerous and less defensible sectors, strong-point strengths could be increased to company size, the distances in Figure 1 closed, and additional divisions could be brought up for deepening the defence.

The defender's problem in this scheme is to identify and concentrate around the attacker's penetration efforts. The Warsaw Pact does not have the combat forces and logistic support to attack in strength everywhere across the front. It will most likely try to give the appearance of strength across the front – but the strength is not there. NATO's problem is therefore one of a progressive elimination of penetration candidates to no more than a half-dozen and more likely to only two or three major efforts. Surveillance of rear areas to determine where enemy reserve divisions are assembling is one method of determination. Nevertheless, because of deception, surprise or simply an enemy reacting to circumstances, defending commanders must plan a system whereby most of the front is screened and held by specially designed economy-of-force units, while local operational reserves are deployed only when forward units appear insufficient. When committed local reserves also appear insufficient, addi-tional reserves must be deployed, using the time

[72] In peacetime the forward divisions would necessarily have to be Category I counter-attack divisions to guard against surprise attacks after mobilization began. These divisions would be replaced by simpler defensive divisions held in Category II readiness.

gained by the blocking action of the previous echelons, the construction of obstacles (but not elaborate barriers) and the harassing effect of night-delivered anti-vehicular minelets on enemy access routes. Because the enemy's strength is itself limited, this technique progressively eliminates the penetration candidates. The attacker can push through several layers of deployed reserves only if he himself is strong; and only major penetration efforts have that strength. When developed, these major penetrations may have a breadth of 30-40km.

After the major penetration attempts have been identified, NATO would want to minimize its own activity across the remainder of the front and reconstitute previously committed forces into local reserves. Unless NATO is planning a counter-thrust in secondary sectors, combat activity in them has little bearing on the overall outcome; it makes no difference if the enemy's secondary forces are badly mauled if in the interim his main effort penetrates and envelopes the defender's rear. The defender's interest in the secondary sector is only to prevent it becoming a penetration sector later, either as a delayed pincer movement or because of blockage elsewhere.

Combat activity in secondary sectors should therefore be minimized to avoid wasting logistic resources. Indeed, under the restructuring proposal of this paper, combat activity *must* be minimized because the logistic resources would no longer exist to support sustained combat simultaneously across the whole front.

The purpose behind reconstituting into reserve those forces that were initially used to identify the major penetrations is that chequerboarding the area in depth around the major penetrations would typically drain half or more of the defensive divisions allocated to the general reserve.[73] Given that armoured and tactical nuclear warfare is potentially fast moving, provision must be made to constitute a new general reserve in case the residue of the initial general reserve was suddenly committed to counter or to exploit unexpected events.

The Achilles heel in the chequerboard system is not the system itself; rather the weak link is when the chequerboard cannot be organized, through surprise or from enemy tactics exploiting opportunities. The counter to this is twofold. If a penetration does occur, the armoured counter-attack forces are large enough to conduct a mobile defence until the defensive divisions can stabilize the battlefield with their movement-choking grid. This would mean, however, that territory would have been lost and may not be recoverable, given the new power of organized defences. *Second, tactical airpower can be designed specifically for penetration tactics. The role of airpower in the defence, as is discussed on pp. 37-44 should be primarily that of filling any gap created when ground forces are deploying or when they become disorganized.*

A strongpoint defence does mean wide gaps between adjacent units, but this need not be dangerous. Automatic weapons and improved artillery ammunition can prevent dismounted foot infantry from being a serious threat in the open terrain of Europe. Enemy infantry can infiltrate at night (a favourite World War II Soviet tactic), but such tactics are inherently slow-moving and are really only useful for undermining the defence before a tank thrust or other heavy attack. Viable strongpoints neutralize this tactic; infiltrators must concentrate in order to attack, exposing themselves to countering artillery fire.

Against tanks, the defender's best tactic would be to allow the Soviet Union her own preference for pushing through soft spots or gaps. With present style defences isolated units do not have the anti-tank means to have a significant effect on enemy forces not directly engaged and often panic when cut off, but this need no longer be true. In a strongpoint defence, porous warfare is expected and even desired, as long as the enemy does not penetrate completely through the strongpoint belt. With on-vehicle and on-site supplies, strongpoint forces are not dependent upon immediate resupply.[74] If the enemy does not overrun the defences, the strongpoints can continuously harass and destroy the attacker's vehicles at long range by direct fire anti-tank weapons and by guiding tank-killing precision-guided artillery fired from the flanks and rear.

Gaps can also be screened by the battalion's new anti-tank cavalry company. Gaps only become a problem during poor visibility, when point targets become difficult to hit. But mist and darkness also restrict the attacker's manoeuvrability. Small ambushing teams from the strongpoints, using poor visibility as their cloak, can wreak havoc upon an advancing opponent, who can only neutralize this tactic by using dismounted infantry on the flanks of

---

[73] For example, of the mobilized 80 divisions, 5 might be forward anti-tank cavalry divisions, 15 assigned to forward defence and 25 formed into counter-attack divisions (using existing tank inventories). This would leave 35 defensive divisions in general reserve (less additional cavalry divisions and light infantry divisions with a dual European/Third World orientation). Pages 23-24 indicated readiness levels of 15-35 Category I (depending on the American, British and Canadian contribution), 20 Category II and 25 Category III divisions. The cavalry counter-attack and intervention divisions, because of their more complex nature, would be held in Category I readiness; the remaining divisions composed of a simple defending infantry would be held in Categories II and III.

[74] Fuel and artillery/mortar ammunition comprise about three-quarters of the aggregate tonnage consumed in a division. Positioned strongpoints consume little of either.

his column and with his lead tanks. This process is slow, consumes relatively scarce Soviet infantry, and can be partially neutralized by the defender's artillery (including artillery-fired minelets) targeted on the exposed infantry.

If the attacker is to undermine the strongpoint system by passing through the soft spots (as opposed to assaulting it point by point) a penetration of depth of 10–15km on a 20–30km front is required, so as to remove the most forward strongpoints from under their artillery cover.[75] But this is difficult because the by-passed strongpoints can operate against the forward flow of enemy resources, and some strongpoints sitting astride the attack approaches will require a costly assault against them. The defender's own tank reserves can use the strongpoints as fire bases and pivot points for counterattack. The attacker's best opportunity for penetrating through to the artillery belt is during a prolonged period of poor weather, which normally comes only in winter when cross-country movement is most difficult.

The purpose of the chequerboard defence is to embed attacking armour in a defensive grid, so as to stop its forward movement, weaken it and create the opportunity for counter-attack. Whereas the mobile defence concedes space[76] and generally forgoes the advantages of static defences in order to neutralize the attacker's advantage of the initiative and concentration, the chequerboard has the advantages of the mobile defence without its disadvantages. By countering possible penetrations with a growing web of strongpoints, first from local reserves, and later from general reserves and reserves reconstituted from other sectors, the chequerboard defence can (a) neutralize some of the attacker's traditional advantages of the initiative, concentration and choice of the point of attack; (b) obtain the defender's traditional advantage of 3 to 1 at the local defence level;[77] (c) defend forward in all sectors except those of the major penetration thrusts, where some space must be traded off to absorb the momentum of the attack; (d) rely more upon simpler defensive forces and less upon large and expensive tank reserves.

---

[75] When the new long-range artillery technology is adopted, the attacker will have to penetrate twice as far and wide. Parenthetically it should be noted that the attacker is more vulnerable to artillery (and tactical airpower) than is a dug-in defender. The attacker's artillery (and air defence) mainly suppresses the defender's fire, so the attacker's manoeuvre forces can close.

[76] Often overlooked is that withdrawals are a major cause of tank losses. Tanks have a high breakdown rate. If space is given up too fast, withdrawing units may be unable to retrieve temporarily disabled tracked vehicles.

[77] Because of the inherent advantages to the defence of such new technologies as terminally-guided warheads, immobilizing minelets and artillery bomblets, the entrenched defender's advantage of 3 to 1 will increase.

In short, a chequerboard is the preferable form of defence against armoured attacks and indeed allows the defence to dominate the battlefield. A mobile defence, on the other hand, has little inherent advantage over the offence because it basically seeks to counter the attacker with the same type of tactic – a battle of manoeuvre. The chequerboard suppresses movement. The artillery shell and the machine gun once did this to infantry; the proliferated anti-tank weapon and the immobilizing minelet allow the defence to do the same to the tank.

Tactical nuclear weapons: chequerboard or mobile defence?

As discussed earlier, NATO's problem with tactical nuclear weapons has been insufficient conventional forces plus some doubt about the political will to use the weapons when the time came. While tactical nuclear weapons inherently ought to favour the defence in Europe, they have in fact favoured the Warsaw Pact. The reason is simply that, with a defence lacking depth and composed of a relatively small number of large battalions, it is relatively easy for the attacker to identify these battalions, attack them with nuclear fire and pour armour through the resulting gap.

Restructuring for larger combat forces redresses these deficiencies, and a mobile defence becomes feasible. It puts the defender on essentially the same terms as the attacker: the greater vulnerability of the defender's forward canalizing defences is offset by the attacker's greater vulnerability thereafter. A chequerboard is more advantageous; vulnerability of the forward forces is reduced by dispersion and smaller units, and the defender's advantage in target acquisition is considerably enhanced from observation by the many strongpoints. Once the chequerboard is established, blasting through it is only feasible by covering the whole terrain, using large yield weapons. (See Table A on p. 31.) However, once contact is made between opposing ground forces, large-yield nuclear attack in the forward areas becomes less likely because of the danger to the attacker's own troops – and few armies are willing to sacrifice their own troops lightly. Moreover, if such weapons are to be used, ground forces are no longer so important – the war would quickly degenerate into mutual destruction.

But if nuclear weapons are restricted to those of low yield, the chequerboard makes nuclear weapons more attractive to the defence than to the offence. The essence of tactical nuclear warfighting is to keep the enemy more concentrated and vulnerable to nuclear weapons than oneself, while avoiding conventional defeat in detail by being too dispersed. A chequerboard, small-unit defence meets both these criteria; mobile defence does not meet the first

Table A: Weapon Effects Table (surface bursts)

Weapon yield in kilotons	Moderate damage to parked helicopters	Prompt casualties to exposed personnel	Prompt casualties to protected personnel	Moderate damage to wheeled vehicles	Moderate damage to tanks; severe damage to artillery
			DISTANCES IN METRES*		
0·5	710	470	330	190	80
1	930	570	420	260	130
10	2,170	930	730	730	350
100	4,960	2,570	1,130	1,970	950
1,000	10,690	7,110	2,200	4,940	2,390

* The table indicates the probable minimum radius of destruction for an expected fractional damage of 0·9 to an accurately located circular area target having uniformly distributed elements.

SOURCE: *Staff Officers' Field Manual: Nuclear Weapons Employment Effects Data*, FM 101-31-3, Departments of the Army and the Navy, February 1968.

criterion. Blasting the many strongpoints in the pre-attack phase is not attractive because of the large numbers of them within a penetration path, their ease of replacement and the need and difficulty of identifying them, particularly with a strong defensive cavalry screen and the deception inherent in small size. Blasting them after contact has been made is possible with the new small-yield nuclear weapons; but unless troop-safety precautions are forgone these precautions can lead to time delays and confusions, increasing the attacker's own vulnerability to counter-attack or nuclear weapons. Finally, to maintain momentum in the new environment of proliferated anti-tank weapons, the attacker must overwhelm the defending system of variable size strongpoints and screening cavalry with more than a 3 to 1 local advantage, as well as pushing into the gaps and the next layer of strongpoints. Yet the very presence of nuclear weapons ensures that the attacker cannot concentrate too much to overcome the layered strongpoints by conventional attack, without becoming overly vulnerable to nuclear weapons.

The restoring function
Warding off the unexpected and restoring the cohesion of the defence is the third defensive function (see p.18). The resources devoted to this obviously depend on the capabilities of the two other functions – those of attrition and holding. If either of these functions is relatively dominant, as in World War I, rearward reserves can be small. If the front cannot be easily held and warfare becomes fluid, strong reserves are required. Lack of reserves is presently NATO's great shortcoming, although reserves would be available after many months of extensive mobilization and sea reinforcements from North America.

Restructuring would provide the reserves necessary for a viable mobile defence, but restructuring and

new technology also provide the means for a chequerboard defence. As far as reserves are concerned, the difference between the two tactics is the relative size of the tank reserve. The mobile defence stresses the tank counter-attack; the chequerboard stresses concentrating defensive forces around possible penetrations.

A chequerboard should contain enemy penetrations and should overcome the volatility inherent in armoured warfare and tactical nuclear warfighting. But, as Clausewitz expressed in his term 'fog of war', the expected is likely to become the unexpected, and a chequerboard defence also needs a strong tank reserve to destroy an enemy weakened in a successful defence or to cover a withdrawal in case of failure. These are, of course, standard missions for reserves, but there are three important points of difference for tank reserves in a chequerboard defence as opposed to current NATO practice for counter-attack forces.

First, the counter-attack reserve for a chequerboard need only have a strike mission and need not consist of combined arms teams capable of defence and offence: the defence can be performed by the specialized, more cheaply equipped strongpoint forces. Ground holding and defence consolidating elements can be omitted, leaving the counter-attack force heavy in tanks as compared to infantry.[78]

Second, strike forces need little organic artillery or logistic support. They will either be in reserve or attacking in or around the chequerboard, the area where logistic and artillery assets are to be concentrated in support frameworks. When committed, the strike force will receive priority of effort and can

---

[78] In the few instances where the tank-heavy force might require more infantry, it can be made quickly available from centralized pools of helicopter-carried light infantry – forces the United States has a surplus of, due to her non-European commitments.

'plug' into both the logistic and the artillery frameworks, consequently having as much as 40 per cent of its total strength in combat platoons (as compared with 24 per cent in present American divisions).

Finally this revision removes a dangerous ambiguity in NATO's doctrine of the mobile defence. That doctrine considers the division as the smallest force capable of a mobile defence and then proceeds to place considerable stress upon the division itself conducting a mobile defence. A division in Europe is not large enough to conduct a mobile defence on its own by splitting itself between canalizing forward forces and a counter-attack reserve. This violates the doctrinal canon that a counter-attack should not be launched until the attacker has been stopped or at least significantly slowed. Because the Soviet Union stacks up large numbers of divisions in the breakthrough area in order to swamp the defence and to exploit the breakthrough, a division's forward forces cannot possibly contain the main breakthrough thrusts, and its reserve can only risk a frontal attack against the penetration. If a flanking attack were attempted, the divisional reserve would find itself subjected to a counter counter-attack on its own flank by a much larger enemy reserve waiting for the breakthrough. Yet if the counter-attack is limited to a frontal attack, no advantage is gained from manoeuvre and the mobile defence. The local 3 to 1 advantage of being on the defence is forfeited against a much larger enemy force.[79] It is therefore better for a divisional reserve to adopt a more passive tactic of occupying a blocking position.

In restructuring and providing for a chequerboard defence this doctrinal ambiguity is removed. More divisions are provided, and divisions are downgraded as entities. Divisions would now be regarded as specialized components of a larger scheme: some divisions would be specialized for holding terrain, others for striking, as in the German Army in World War II.

The attrition function
Seriously weakening an opponent before engaging him in close combat is a standard military tactic and, throughout history, missiles have usually been exchanged before the main bodies have clashed. With few exceptions this type of exchange of fire has been only marginally effective, but by the mid-1980s attrition at long-range may become a significant factor. Precision-guided warheads are already beginning to suggest its possibilities. Their development should further favour the defence.

Guerrillas, ground-screening forces, artillery and tactical air power are the present means of attrition forward of the main line of defence. Each has major-limitations; with present technology none is particularly effective without nuclear weapons. A guerrilla system requires considerable time to establish. Long-range and stay-behind patrols can be organized immediately, but have been designed mainly for reconnaissance and only secondarily for such combat tasks as harassing the enemy, tying down his reserves and, during major battles, disrupting communications behind critical sectors. In order to survive, such groups have to use hit-and-run tactics, concentrating on demolitions, destruction and fast dispersal. Precision-guided weapons simplify their task. Rather than elaborately planned surprise assaults (by definition limited in number), precision weapons offer the prospect of the missile delivered from a distance, with a circular error probable (CEP) measured in metres. The stay-behind patrol now need only communicate the details of the target to a control system in friendly territory. Their more conspicuous equipment can be limited to a transmitter and receiver for exchanging information about stationary targets; for moving targets or for for achieving even greater accuracy against static targets they can carry a radio and laser designator packaged like a camera. Thus, whereas in the past guerrillas and stay-behind patrols emphasized reconnaissance, because of the high risk associated with overt action, new stand-off technology permits their combining stealth with discriminate destruction.

Ground screening and general outpost forces are primarily for early warning and delaying the enemy; their combat power and attrition capability has been limited. Their advantage is that they have been essentially 'free'; when withdrawn, they become part of the main body. Such forces can now have a good anti-tank bite, but technology cannot remove their second limitation of being only a temporary attrition force; they are not strong enough to hold and defend; when withdrawn, their attrition role ceases.

The limitations of artillery have been imposed by difficulties of target acquisition, range and accuracy against point targets. New technology can ease these constraints. Replacing howitzers with longer-tubed guns of higher muzzle velocities (as in the Soviet Army) can double ranges to 30km; rocket-assisted propulsion can extend it as far as 40km (as against 15km for standard 155mm and 8 in. howitzers). Extended range gives proportionately greater depth into enemy territory because the gun need only be

---

[79] This is probably an example of a principle of war causing a doctrinal blind-spot. According to these principles, only the *offensive* can be decisive. This is impressed upon all officers and ranks. It then becomes an article of faith that morale will collapse without offensive action – hence the well-known phenomenon that, though war plans may call for a withdrawal, armies largely train for the attack. In NATO's case a contributory reason for this stress on mobile defence at division level is that the scarcity of divisions has caused the division to assume unwarranted significance.

placed the same distance behind friendly lines as the shorter range howitzer. For even greater range, there is of course the more expensive missile. Most important, artillery with cannon-launched guided projectiles can be made effective in good weather against both stationary and moving point targets.

Finally, 'smart' bombs greatly increase the effectiveness of tactical air power. Without guided munitions the accuracy of air-delivered ordnance has become so poor as to downgrade the importance of tactical air power. While guided munitions are more expensive than conventional bombs, they increase kill probabilities more than a hundredfold for point targets like tanks and bridge abutments, while marginally reducing aircraft losses.[80] Unfortunately for tactical air power, however, technology is a double-edged sword: it also introduces more effective air defence and new and cheaper ways of dealing with interdiction for ground support targets.

If it were not for the increasingly inhibiting effect of air defence upon tactical air power, a new dimension in warfare capabilities might already be at hand. Targets – provided they can be identified and accurately located – can now be destroyed regardless of range. The problem is target acquisition. The stay-behind party behind enemy lines and the aircraft pilot represent the only means now available for finding non-topographical targets accurately and in time, so as to make the precision of the warhead

usable. It is difficult to use guerrillas or stay-behind parties in rapidly shifting situations, and the strength of air defence is such that manned aircraft may suffer heavy losses; their use may no longer be cost-effective for conventional interdiction targets. The new dimension of attrition warfare therefore depends on remote target acquisition, coupled with a reduction in the cost of terminal guidance to point targets when laser designation from ground observers is not feasible. This cost is at present often prohibitive.

Stand-off technology[81]

The major cost element is the guidance system, determined by the type of target acquisition and whether range is such that both mid-course correction and terminal guidance are needed. For the purpose of estimating approximate costs, targets can be classified into an increasing cost order of those which can be acquired by ground observers, those which are known and stationary, targets emitting an electronic signal, and moving targets which cannot be observed from the ground.

Observed targets can now be destroyed at costs of only several thousand dollars if the target can be laser-designated and the warhead accurately pointed into the conical area (or 'basket') illuminated by the laser. These low costs are limited, however, to artillery gun and missile ranges and manned-aircraft-

---

[80] The cost of a 3,000 lb 'dumb' bomb is said to be $1,800; a laser-guided version costs $4,900; a TV (electro-optical) version of the same bomb might cost some $15,000.

The cost-effectiveness potential of 'smart' bombs lies in the fact that their kill probabilities are very much higher, the ratio depending on the radius of the area within which any bomb must fall to destroy the target, and the relative delivery accuracy. Assuming a spherical normal bomb pattern and weapon reliability equivalent to 'dumb' bombs, the greater effectiveness of the 'smart' bomb is shown below.

Circular error probable (CEP)		Radius within which bomb must fall to destroy target (ft)	'Smart' bomb times more effective than 'dumb' bomb
'Dumb' bomb (ft)	'Smart' bomb (ft)		
750	10	10	× 4,065
		20	× 1,900
		30	× 900
300	10	10	× 645
		20	× 300
		30	× 145

If the bombs are used against less concentrated but vulnerable targets, such as infantry, the relative advantage of the 'smart' bomb would go down, but against hard targets within a small area they are – at the least – 145 times as effective, as the table shows (the figure of 300 feet for the CEP of a 'dumb' bomb is based on practical results in Vietnam in 1971).

Additional advantages of 'smart' weapons are that aircraft can drop them from higher altitudes and from greater

ranges, and that the aircraft are not locked into prescribed flight paths, and so can take evasive action during the delivery run without affecting accuracy. This means less risk to the aircraft and thus a lower attrition rate. The extra cost of 'smart' bombs – quite apart from their extra effectiveness – is cancelled out if the attrition rate is only about ½ per cent per sortie less. Laser delivery reduces attrition, since large bomb delivery tolerances are possible and the vulnerable period of line-up for weapon delivery and computation prior to bomb release can be reduced.

[81] Many of the possibilities discussed in this section have not yet been engineered or tested, but the system components are generally available from present technology or require only moderate development. Many of the militarily more important – and less technically glamorous – possibilities could be fielded within several years if they were seen to have utility, so that priority could be given in resources and management attention, and the systems taken out of the normal development process.

delivered laser-guided weapons (exclusive of sortie attrition and operating costs). 'Lock-on' terminal-guidance (e.g. electro-optical and imaging infra-red) from manned aircraft increases costs to about $15,000. For longer ranges, with ground-observed laser-guided weapons, mid-course correction may be required to guide the missile to the area of laser-illumination; this increases warhead costs to $30–50,000.

If no observer is available, stationary targets can be obtained by surveillance (by satellite, for example) or, for topographic features, by using map co-ordinates transferred to a reference grid, the more promising involving the measurement of propagation time of electromagnetic radiation from two or more stations. The CEP of the various mid-course guidance systems[82] is generally of the order of 100 feet or more, but can be as low as 25 feet for a moderate-range (50–100km) co-ordinate grid, using ground-based distance measuring equipment (DME). Besides giving greater accuracy, ground-based DME needs only relatively cheap ground stations, two per guidance, costing about $100,000 each. Airborne DME has greater range and operational flexibility, but is less accurate (ground stations can be very precisely located) and its stations are extremely expensive (several hundred million dollars for a grid the size of central Europe).[83] A disadvantage of DME is that it is an active, two-way ranging system, but this disadvantage can be minimized as is discussed on pp. 38-39.

Similar accuracy can also be obtained from active one-way ranging and the passive reception and calibrated clocking of received signals. This system virtually removes the vulnerability of the in-coming missile, but its airborne and space guidance platforms are vulnerable and very expensive (upwards of $1 billion for a large grid). Such a precise and centralized position-fixing system also offers a number of collateral benefits – a common grid for command and control, the reduction in operating costs by phasing out other less accurate systems, and the possible integration of a high-accuracy grid system with local communications, navigation and identification systems.

With large 1,000–2,000 pound warheads, reference-guidance systems are accurate enough for area fire using bomblets, and even for point targets for the relatively small-CEP ground-based DME system. For

stationary targets requiring pinpoint accuracy an additional terminal sensor is required. If the target is a continuous active infra-red or electronic emitter, the terminal sensor can be inexpensive. Otherwise sophisticated imaging sensors and a secure data link for command and control are also required. This more than doubles the cost into the $100,000-plus category for weapons systems like the American *Condor* and *Harpoon*, limiting their usefulness to high-value targets.

Targets emitting a strong discontinuous electronic signal, like air-defence radars, can be acquired by time-of-arrival electronic triangulation techniques, and then targeted like an unobserved stationary target. Technology for unobserved *moving* targets is complex and likely to be impractical for anything other than very short ranges, except for moving-target-indicator (MTI) radars and command guidance. Not only must the target be found, but it may continue to move during the preparation for firing. This implies an expensive search process added to the high costs of hitting a remote target.

For missiles with a very short range of a few kilometres, search and data-link problems may now be solvable by techniques analogous to the command-guidance of wire-linked ATGM, at a cost of only several thousand dollars. Observation can be obtained by an imaging camera (electro-optical or imaging infra-red) in the missile nose. Sensor cost can be cheap; low resolution is sufficient, since the target's approximate location must be known and because of new electronic developments like the charge-coupled device. The critical element, however, is the high-capacity data link between the imaging camera and the observer; this now appears solvable by fibre optics (a light, high-tensile, high-capacity cord). This system gives the firer a limited high-angle search, sufficient for hitting defiladed targets.

A less elegant solution for attacking armour in defilade is the heavy mortar with a passive infra-red seeker.[84] Its advantages are that it is not too limited by weather and that it has a high rate of fire. Its disadvantages compared with the fibre-optics system are its susceptibility to spoofing when the enemy is so disposed, and the imaging camera's confirmation of the enemy's approximate location after the first round. However, much more important – though less glamorous than the stand-off capability – is the mortar's capability with an infra-red seeker against assaulting tanks on the battlefield itself. Since in this case the enemy's location would be known and the angle of descent would be nearly vertical, the infra-red seeker needs only a narrow field of view, and the

[82] Examples are *Omega, Loran*, time of arrival, DME, calibrated position fixing, inertial, automet, and area correlation (a programmed map-matching technique). Their purpose is to guide a projectile by mid-course correction to the proper area for dispersing area bomblets or into a 'basket' suitable for the terminal-guidance sensors.

[83] For a discussion of electronic counter-measures and defence suppression see p. 40 *et seq*. Airborne DME can be simplified by the techniques discussed on p. 32 for MTI radars.

[84] Infra-red is primarily a night sensor, with some capability for penetrating light-haze and smoke. Microwave radiometry, which is similar to infra-red, is uninhibited by clouds, rain and smoke.

guidance package can be simplified and more cheaply designed (i.e. the seeker can be 'strapped' eliminating costly gimbals) for about $1,000. The important implication is that Soviet swamping attacks could be countered by saturating the area with cheap terminally-guided warheads without the assistance of an observer and often without regard to visibility conditions in the defensive position. Moreover, with appropriate sensor filters, spoofing would be difficult in this case.

For unobserved moving targets at longer range, beyond that at which a fibre-optic is practicable, three solutions may be possible: a 'lock-on' imaging camera using a line-of-sight laser data-link, remotely piloted vehicles (RPV) and MTI radars with command guidance. While limited to use in fair weather, a laser data-link ranging out to 50km is less costly than other secure data-links. 'Lock-on' imaging circuitry, as in the air-delivered *Maverick*, has been complex, but promises to become smaller and cheaper, using charge-coupled devices Thus in an area where the enemy's location is known (e.g. a main supply route or targets identified by real-time surveillance) and at ranges where the accuracy of the rocket is acceptable, a *Maverick*-type attack can be mounted with free-flight high-angle rockets, costing only slightly more than the air-delivered *Maverick*. Weather conditions would, of course, be more restrictive than for air delivery, but expensive aircraft losses would be avoided and the rate of fire, when usable, would be higher.

For RPV a number of modes are possible, the general problems being their survivability and the complex command and control needed. For unobserved moving targets all require imaging sensors and data-links. Whereas these requirements have led to high costs in other stand-off systems, the RPV attempts to reduce costs by a combination of simple, toy-like construction, close-in imaging sensors requiring less resolution, simple data-links, recoverable airframes and target designation for laser-guided warheads or for RPV-mounted *Maverick*-type missiles. These combinations lead to three broad groupings:

(1) An austere RPV with a small shaped-charge warhead with a range up to about 10km and a costing goal of about $10,000. While small, it could be easily hit by ground air-defence fire and readily jammed (in which case it would move on to another target). The system's utility is likely to be bee-like: complicating the attacker's environment with a swarm of cheap flying bombs. An alternative form of the austere RPV is a surveillance drone for area fire weapons. When targets are located, the drone's location is electronically determined, and artillery is fired into the area.

(2) Remotely piloted aerial observation and designation systems (RPAODS), ranging out to 15–20km with a costing goal of $75–100,000. These recoverable platforms would designate targets for laser-guided artillery and would be the most complicated of the RPV systems. Problems are co-ordination between the designator and the firing weapon, determining accurately enough the target location from the platform's position and the need for the platform to take evasive movement during the laser designating period in order to survive. The data-link must also be secure.

(3) Unmanned aircraft carrying *Maverick*-type missiles, with a range of 100–200km and a costing goal of $200–300,000. This system simplifies the command and control, location and survivability problems inherent in the RPAODS approach, but its airframe and missiles are significantly more expensive, limiting attractiveness to airframe attrition rates of 20 per cent or less for most target categories.

The third solution for attacking unobserved moving targets, that of MTI radars, seems the most promising, with costs eventually as low as $30,000. Its basic attraction is combining surveillance and real-time command guidance into a single co-ordinate system. This enhances accuracy, rate of fire and command and control. Potential limitations are electronic counter-measures and platform vulnerability, particularly to emissions-homing missiles. Wide frequency spectrums and random frequency selections are, however, making radars increasingly difficult to jam. The cost of platform vulnerability can also be significantly reduced by separating the radar and the carrying platform: electronic pods can be suspended from high-flying helicopters and balloons, which as a side effect enhances radar resolution relative to high speed platforms.

Military dividends

Technical feasibility and military desirability are not synonymous, and relative cost-effectiveness to perform a given mission (so-called mission analysis) is an insufficient criterion. Stand-off attrition weapons are already competitive with tactical air power when the air-defence environment is a hostile and sophisticated one. The interesting questions concern costs and how these weapons might affect the battle on the ground. If the weapons are expensive – and many will undoubtedly remain so – their military value depends entirely upon whether or not they have an important effect upon the battle, not in their cost-effectiveness as related to tactical air power.

At present costs, unobserved stand-off weapons are suitable only against high value targets, mainly key transport nodes, major command and control centres, air defences and air bases. The last two categories provide the bulk of present potential targets, since they are considered by NATO staffs as critical to the problem of obtaining air superiority, so as to release air power to support the ground forces. But it may not be easy to destroy enemy aircraft, because they might be in shelters or be dispersed. The enemy air-defence system may have heavily protected and overlapping control centres, and its weapons may be difficult to avoid, jam or suppress (see p. 40 *et seq.*). In such a case, attacking these targets and destroying or suppressing them with high-cost weapons may not be warranted, or be given the priority that it may be now.

If unobserved stand-off weapons become considerably cheaper – say about a third of present costs – they can then accomplish two of tactical air power's coveted (and partially forfeited) missions – long-range supply interdiction and battlefield interdiction. The purpose of the first of these is to restrict the flow of supplies; the second is to isolate the immediate battlefield from its source of supply and most importantly to prevent the movement and deployment of reserves (particularly at the time of a breakthrough).

A cheap stand-off weapon which could be targeted on known locations could have dramatic implications: movement in enemy rear areas could be so restricted by knocking out the transport system that large-scale attacks would become impractical, *provided the attacker could be held long enough for the effects of interdiction to be felt.* With such a weapon[85] the defender's primary concern is reduced to that of having enough strength to hold against surprise attacks. The transport grid in the battlefield area could not support concentrated armoured formations for breakthrough and exploitation tactics. The alternative then open to the enemy – an across-the-front style of attack rather than a narrow thrust – is feasible but unattractive. It would increase reliance upon firepower and so increase total logistic demands. With the manoeuvre of forces limited, it would be difficult to exploit success, and the time between attacks would lengthen as artillery ammunition was slowly brought forward, stored and rapidly consumed in major attacks. At best the result would be an unrewarding series of costly frontal attacks against successive prepared defences, which would not give the Warsaw Pact the quick and decisive victory it needs. NATO's superior resources could be brought to bear and the prospective costs of protracted conflict could induce either a negotiated settlement or nuclear escalation.

Of the technologies concerned with moving targets, the technically exciting RPV capability is militarily less interesting than the heavy mortar with

an infra-red homing sensor. The role of RPV and MTI radar is battlefield interdiction, but, for countering penetration tactics, isolating the battlefield and disrupting the movement of enemy reserves are more effectively accomplished by destroying the transport net and physically sealing off the penetration area than by attempting to wear down the enemy by attacking moving targets (which might be replaced promptly from the enemy's large reserves). Furthermore few moving targets warrant destruction by a high-cost system; enemy command vehicles and tanks in a breakthrough sector are among the exceptions. Defence against tanks can now be handled by the new family of cheap anti-tank weapons and, as it becomes operable, 'smart' artillery; especially in a chequerboard defence presenting many fronts to an enemy tactic attempting to move rapidly into the defender's rear. Finally, destroying moving targets through locating them with sensors is inhibited in the breakthrough sector by the resolution required to distinguish between interspersed friendly and enemy forces, and by European terrain characteristics in general. The older German roads are tree-lined, half of Germany is forested, and all armies use forests for assembly areas and attack positions, often moving into them during darkness. Trees have adverse effects on terminal-guidance weapons. Besides often rendering tank-destroying munitions ineffective, they provide camouflage, reduce the contrast needed for electro-optical guidance, and are considerably more reflective than military targets to laser beams.

While the remarks above have tended to emphasize its limitations, technology that will enhance attrition warfare is now at hand and, as it matures, costs should drop. To be really useful, attrition must be linked to an ability to defend *forward* and to prevent enemy occupation of friendly territory; this also gains the time necessary for attrition to take its toll.

The new attrition weapons have three major politico-military implications. First, when combined with a strong holding capability they can neutralize a *blitzkrieg* style of attack, substituting instead a less dangerous but protracted infantry-style conflict. Second, attrition technology inherently favours the defence, since the attacker must expose himself in movement and must use a developed road net if speed and decisive results are desired. The defender, on the other hand, can conceal himself in the forested hills and stone villages astride the road net in Germany and supply himself from small, pre-stocked forward depots. Finally, the impact of the new modes of attrition may be institutionally wrenching upon the military services and will challenge the division of roles and missions between the services, particularly affecting tactical air power.

---

[85] An attraction of such a system is also that it need not be time-urgent.

Doctrine and technology should interact. Doctrine is constrained by technology, but doctrine also indicates which technologies to pursue. Sometimes this interaction is best served by increasing military capabilities, if only because the enemy is increasing his. Sometimes, however, technologically induced reorganization may produce redundancy and hence cost reductions. Technology should not be thought of as just a means for improving and replacing old weapons with new, obtaining greater firepower, or shoring up 'experience-proven' tactical concepts. Just as technology can be used dialectically to improve effectiveness, so it can be used to reduce costs through higher equipment productivity, the simplification of certain functions or their outright reduction where they are made less relevant by technological progress. The number of possible examples is clearly large. This chapter discusses major cost savings in two of the most expensive military functions: indirect fire support by artillery, and tactical air power.

Artillery

In modern European armies, artillery (including mortars) is often the largest branch; it also makes the largest logistic demands. In the United States Army, for example, artillery and mortar ammunition is likely to account for a third of the tonnage flowing into a divisional sector.

New technology affects artillery in several ways. Terminal-guidance enhances its role and much reduces the need to augment it with air-delivered ordnance. It gives artillery a fair weather capability of destroying tanks, *in addition* to its undiminished role of area fire, thus adding an important new dimension to artillery at virtually zero incremental cost.[86] This itself does not imply major reductions in the number of guns needed, because, even though fewer shells are necessary for many types of target, the artillery's peak load capacity needs to be maintained for bad weather conditions and for occasions calling for mass fire.[87] For real savings to be made, technology has to give a more generalized increase

in productivity, so that greater effectiveness can be traded for cuts in numbers.

As far as accuracy is concerned, the three major types of targets are: point targets requiring great accuracy; small targets requiring relative accuracy; and saturation fire, where accuracy for individual weapons is less important. In NATO, tube artillery is generally used against all three types. This is obviously inefficient; tube artillery, which is expensive, is really only suitable for the second category of small targets. Against point targets it is not particularly effective; against saturation targets it is unnecessarily expensive. A more cost-effective solution for the present would be a mixture of tube artillery, which would be used for relatively accurate fire across the front, and of rocket artillery, which is cheaper and has a high rate of fire, for saturation fire. Several NATO countries, taking a leaf from the Soviet book, are now organizing their artillery in this way.

New technology, however, can provide a still more cost-effective answer. With terminal-guidance, rocket artillery – previously only suitable for saturation fire – is now suitable for point and small targets. For point targets terminal-guidance is required for both tube and rocket artillery. For small targets terminal-guidance is often preferable, even for tube artillery, the greater cost of terminal-guidance being partially offset by the fewer rounds needed and greater effectiveness obtained by rounds arriving on the target without the warning that is given by present ranging methods.

Even more cost-effective, however, are the possibilities suggested by rapid-fire gunnery and extended range. Present howitzers have rates of fire of around three rounds per minute for short periods and one round per minute for sustained fire,[88] but new automatic loading, improved recoil systems and heat removal techniques can allow tube artillery to fire at sustained rates of upwards of 30 rounds per minute, the limiting factor being reloading the gun magazine.[89] The range of NATO artillery can be doubled from the present 15 to 30km by replacing howitzers with guns using longer tubes and improved propellants giving higher muzzle velocities. Rocket-assisted propulsion (RAP) can further increase this range to 40km. Greater range, of course, has its penalties. A mortar or howitzer shell is more effective than that of a gun because it does not need such a thick casing (to resist firing pressures but

---

[86] All-weather capabilities free from the problems of the forward observer on the immediate battlefield can be obtained from heavy mortars with infra-red/radiometric seekers and from large rockets and boosted glide bombs dispersing anti-tank bomblets and minelets over selected areas. At shorter ranges these rockets can be unguided and cheap; for longer ranges more expensive mid-course correction guidance would be required.

[87] Artillery is often criticized for its inability to stop armour. The criticism, however, neglects the fact that land warfare is very much like the child's game of paper, rock, scissors. Artillery stops infantry and suppresses infantry defensive fire – a particularly important function as infantry adopts the new proliferated anti-tank weapons.

[88] *Field Artillery Reference Data*, United States Army Artillery School, Fort Sill, Oklahoma, vol. II, 1970, p. 231.

[89] The new French 155mm self-propelled (SP) gun, which has some of these features, has a range of 23·5km and can fire up to 8 rounds per minute until its internal magazine of 42 rounds is empty.

which reduces payload) and has a more vertical angle of impact. (This objection, however, is only inherently valid for longer ranges; for mortar and howitzer ranges, guns can fire special shells with thinner casings and at high angles. The penalty is a larger, more mixed ammunition inventory.) RAP has not been considered attractive because of greater miss distance due to range, lower shell payload due to RAP chamber volume, and its cost of several hundred dollars, effectively doubling shell cost. Terminal-guidance nullifies these objections for point targets like tanks and gives artillery much greater mutual reinforcement.

Artillery is of course vulnerable to this accuracy itself, and, as counter-battery technology is developed, future artillery may have to 'shoot and scoot' for survival.[90] Rocket artillery can deliver a heavy weight of fire very quickly and then move. Neither towed nor self-propelled tube artillery has the necessary rate of fire to match the multiple rocket launcher unless rapid-fire techniques, as developed in naval gunnery, are adopted. And towed artillery is inherently vulnerable unless it does move.[91] Rapid-fire guns mean that tube artillery can produce the same throw-weight in the same time as the rocket launcher but without its inaccuracy and propulsion inefficiency, and therefore tube artillery has relatively low ammunition costs and lower transport demands. Rapid-fire guns, however, may be bulkier and therefore occasionally slower to come into action than rocket artillery.[92] Bulkiness, however, can be partially reduced by having a smaller enclosed ammunition magazine and relying upon armoured slave vehicles for ammunition resupply (using fast transfer techniques).

Even *if* rapid fire guns were to prove less mobile and slower into action than rocket or present artillery, this should not really be an inhibiting factor; the criterion is not the responsiveness of single weapons but of the artillery system as a whole. Leap-frogging weapons, together with longer range, means that the artillery *system* could still respond to frontline demands even with slow emplacement weapons. If it were considered necessary to have high mobility for special situations, rocket artillery could be provided, or even a few of the present self-propelled tube battalions could be retained. But this special requirement for mobility should not be allowed to drive the whole system into an inappropriate structure and needlessly high costs.

Through the adoption of rockets or rapid-fire gunnery, artillery productivity in the firing battery could be considerably increased. Overall reductions however will be much less than range increases and improvements in rates of fire might suggest, because the scope for reductions is smaller in the two remaining components of the artillery system – fire control and ammunition resupply – and because the numbers of artillery units (as opposed to their size) cannot be reduced sharply because of an inability to insure against the risks inherent in small numbers. These considerations indicate that in converting from howitzers to rapid-fire guns (1) the number of battalions might be reduced by a quarter, reflecting greater range and capability for mutual reinforcement; and (2) firing tubes per battalion might be reduced from 18 to as few as 3, reflecting an even greater increase in rates of fire.[93]

While the bulk of artillery is conventional, nuclear missile systems form about 20 per cent of total artillery battalion strength. First generation systems – like the American *Honest John* – were relatively cheap unguided weapons, but their poor accuracy, slow rate of fire and general unwieldiness made a conventional role for them unreal. Second generation missiles – like *Lance* – are smaller and have significantly improved accuracy, but their rate of fire remains low and costs have become prohibitively expensive for conventional use because of a reliance upon inertial guidance, or, in the case of *Lance*, a modified inertial thrust-control guidance system. New technology in the form of ground-based DME guidance and boosted glide-weapons, using a semi-ballistic trajectory, can change these limiting characteristics, and missile systems can now be truly dual-capable. They are cheap enough for conventional use and are more suitable for the nuclear role than *Lance* or similar missiles. This technological

---

[90] Counter-battery fire is an example of a technology with a highly leveraged payoff that has remained relatively undeveloped. Just as the use of artillery became of critical importance in World War I to suppress the new automatic weapons of the infantry, proliferated anti-tank weapons suggest artillery will again become critical to suppress anti-tank weapons so as to retain mobility for the tank. Artillery suppression or counter-battery fire (and artillery's responding tactic of 'shoot and scoot') could go far to determine the degree of mobility enjoyed on the battlefield.

[91] The armour on American self-propelled artillery (widely used in NATO) gives limited protection for only part of the firing crew and the fire direction centre; the remainder of the battalion is unprotected. The 8-in. SP howitzer lacks even crew protection. The new French 155mm SP gun has a closed turret and is better protected.

[92] A major constraint on the speed of opening fire has been gun survey. Survey techniques have been labour-intensive and time-consuming, but electronic triangulation, like automatic azimuth referencing, now permits immediate fire for effect for the newer weapons.

[93] These examples are only for illustration. In actual practice such revolutionary technological changes suggest a restructuring of the artillery system. In addition a general tightening of present artillery operating procedures can reduce artillery manning by three-sevenths. For details see S. Canby and R. Rainey, *op. cit.* in note 58, pp. 73–93. These tightening reductions are included in the halving of the divisional slice. Soviet artillery has only half the number of men per tube of American and NATO artillery.

combination can considerably improve accuracy and rate of fire at half the cost of a *Lance*-type missile, while delivering twice the conventional payload for a comparable distance. Launching location and launching azimuth are also not so critical with DME guidance, and supporting equipment can be simplified. This technological combination is unambiguously superior in the conventional role and has but one disadvantage for nuclear weapons delivery – the relatively greater vulnerability of DME guidance to enemy counter-measures. DME can be jammed, but this is unlikely to be an inhibiting factor with fast frequency-hopping and pseudo-random noise within a wide frequency spectrum; in addition, the combination of ground-based station and high-angle missile is difficult to jam because of directional antennae and the terrain masking of the DME station relative to a ground jammer. The more serious problems are enemy interception of the missile and suppression of the DME guidance emitters. As with all electronic emitters, they can be located and homed on. However, as opposed to some expensive radars where redundancy is necessarily limited and data gathering and data processing cannot be separated, several solutions exist. Ground-based DME emitters are relatively cheap ($100,000) and therefore relatively expendable. Their vulnerability can be also significantly reduced by 'hop around' interrogation (i.e. multiple guidance emitters with discontinuous emissions) and wide ground separation of the vulnerable antenna from the DME equipment itself. The missile itself is difficult to intercept by homing upon its radiation, because of frequency-hopping, discontinuous emissions, weak signals, and the general difficulty of a ballistic intercept.

In short, new technology leads to the rather startling conclusion that new anti-tank and interdiction roles can be assigned to artillery while still permitting manning reductions. About 10 per cent of a theatre army force is artillery. Restructuring increases this artillery content to about $11\frac{1}{2}$ per cent (restructuring doubles the numbers of artillery, but at the same time procedures are tightened by three-sevenths). New technological infusions permitting a 25 per cent reduction in numbers of battalions and a reduction in gun tubes from 18 to 3 per battalion (a 25 per cent reduction in residual battalion strengths) lead to an artillery strength of $6\frac{1}{2}$ per cent of the theatre army force, which is a direct saving of 5 per cent of ground and air force manning. Additional savings accrue from the transport-reducing demands of 'smart' ordnance and the multiplied effect of any reduction upon supporting services.[94]

---

[94] Parenthetically, it should be noted that this reduction itself approximates the much-heralded economies from standardization and common procurement and logistics; commonly

Tactical air power

While Germany and Japan were the innovators in tactical air power before World War II, neither developed it to the same extent as the United States. In the Pacific theatre, naval air power was certainly the decisive element; in Europe, tactical air power played the critical role of immobilizing German armour. As a result of this experience and the American desire to use technology to save casualties, tactical air power (navy, air force and marine) has grown into the most costly component of American general purpose forces, having had a slightly larger share in recent budgets than land forces. Aircraft (including army aviation, airlift and strategic bombers) and their associated armament have taken 40 per cent of the total American procurement expenditure. In NATO the United States allocates almost as much for her tactical air forces as she does for her ground forces.[95] For the European Allies air power does not play such a prominent role, but, nevertheless, 20 per cent of European manpower is devoted to air power with, again, a proportionately larger budget share.

With such large resources committed to tactical air power, three fundamental questions must be asked:
(1) What are the roles of tactical air power?
(2) Are the benefits commensurate with the costs?
(3) How should air power be constituted?

The problem
In World War II fighter aircraft could make repeated, low flying passes and expect to survive and to kill personnel and destroy tanks with their machine guns and 20mm cannon. This capability has now been invalidated by changes in ground forces – specifically by better armour and ground air-defence weapons, and also by 'progress' in aircraft speeds – which make line-of-sight aiming impractical except for high angles of descent. Thus 'strafing' ordnance has given way to inherently inaccurate dive-bombing attack at high speed from high altitudes. Improved air defences cause aircraft with ordinary unguided weapons to limit themselves to single passes in dive-bombing attacks. High delivery altitudes require visibility conditions not often obtainable in Europe, thus preventing the use of aircraft; the high performance of aircraft has increased maintenance requirements, which in turn has slowed turn-around time. Finally, these results have been compounded

estimated at $1–2 billion per year – roughly 2–4 per cent of the $50 billion per year spent by the Allies for defence of the central front. See also note 21.
[95] The Brookings Institution, for instance, estimates that 25 per cent of American costs in NATO are for land-based tactical air and 10 per cent for carrier task forces in the Mediterranean (compared with 40 per cent for ground forces). 'Estimated Cost of US General Purpose Forces for NATO and European Contingencies by Region, Fiscal Year 1972', *US Troops in Europe* (Washington: Brookings Institution, 1971), p. 119.

by the economic infeasibility of fielding large numbers of expensive aircraft and the need to divert many of those that there are to supporting missions (air cover, air-defence suppression, electronic counter-measures, rescue, etc.).

Analytical models using the combination of low accuracy, low sortie rates, low aircraft numbers and high aircraft attrition quite plausibly support the army's contention, from experience in Korea and Vietnam, that tactical air power even in a predominantly ground support role has little effect with ordinary weapons on the ground battle. Air-power enthusiasts, on the other hand, attempt to refute this evidence by quoting the examples of World War II and the 1967 Six-Day War. But too many circumstances have changed for World War II experiences to be accepted without critical examination. The empirical evidence of the Six-Day War also gives air power only modified support. Given the weather, terrain, high Israeli sortie rates and complete air superiority, tactical air power should have been in its element. However, of the Arab tanks destroyed, only about 5 per cent showed any signs of being attacked by aircraft and only 2 per cent of being seriously damaged by air-delivered weapons. Its main – indeed battle-winning – impact on the ground battle was in preventing movement of soft-skinned vehicles by day and in blocking columns in confined areas like the Mitla Pass.

The Arab–Israeli war of October 1973 has implications less favourable for air power. It showed that, in a hostile air-defence environment, tactical air power cannot operate without extremely expensive losses. Israel was finally able to destroy the Egyptian air-defence system in one area only, and that by ground forces. This presents a paradox: the purpose of NATO air power is to support ground forces; but if ground forces have to defeat the enemy's ground forces first, what is tactical air power for, except for neutralizing enemy air power?

The conclusion must therefore be that experience since World War II has so far worked rather against tactical air power. The relevant question then becomes: can the adverse combination of low accuracy, low sortie rates, low aircraft numbers and high attrition be overcome? Terminal-guidance of weapons corrects only one element by providing effective ordnance *in good weather conditions* against tanks and other point targets requiring high accuracy. Cheaper and more rugged aircraft like the American A-10 and the Franco–German *Alpha Jet* can give a considerable increase in sortie rates and greater aircraft numbers, but the big question mark is aircraft attrition.[96] If this problem cannot be

[96] Any attrition rate above 2 per cent per sortie is normally considered prohibitively expensive. Compounding a survival rate of less than 98 per cent per sortie means that an air force

resolved, the role of tactical air power in Europe will eventually be reduced to that of a quick reaction force, whose ability is limited to helping the army in situations when it has been thrown off-balance. This is an important, even critical, role; nevertheless, it reduces tactical air power's independent stature.

Unfortunately, it seems that aircraft will remain vulnerable to sophisticated air defence, even if using hyper-expensive electronic counter-measures (ECM) and stand-off suppression weapons. Dense air defence systems with passive and short-range sensors are difficult to avoid, jam or suppress; they are also cheap and their gun components are effective against ground targets. The larger air-defence missiles, which are active emitters and thus susceptible to ECM, are countering this vulnerability with pseudo-random noise and fast frequency-hopping within a wide frequency spectrum. Decoding a hidden waveform or blanketing a wide frequency band with a large power output is likely to be impractical, particularly if corridor blanketing is ineffective and each aircraft must carry its own full range of ECM equipment. Consequently the emphasis has begun to shift from ECM to air-defence suppression. But this is likely to prove similarly unrewarding. Present suppression weapons are too costly for extensive use; their cost-effectiveness is only relative to manned aircraft. If they become cheap they partially replace air power itself. Second, air-defence suppression can be neutralized by the mobility of the ground weapons and by electronic spoofing. Third, automatic anti-aircraft guns are likely to proliferate as a surfeit of anti-tank weapons comes into existence and anti-tank suppression (with such guns) then becomes critical. Fourth, the low-level *Redeye*/SA-7 type surface-to-air missile will become more effective (e.g. have greater speed and laser-beam guidance) and can be made more plentiful and available for artillery, logistic, and even militia units.[97] Countermeasures used in Vietnam, such as flying higher, avoiding air-defence parameters and using faster

ceases to exist after a short period. Attrition estimates, exclusive of the potentialities of 'smart' air defence, already considerably exceed 2 per cent for the expected air defence density in Europe.

[97] The object of field air defence is not fundamentally to destroy enemy aircraft, but to prevent them from interfering with ground operations (the strain upon resources in providing this air-defence effort must not, of course, be unreasonably large).

The important factor for NATO's air defence is whether the Soviet Union will concentrate on air- or ground-delivered 'smart' weapons. Soviet-style tactical air power has so far been designed mainly for air defence, to keep opposing Western ground-support aircraft 'off the back' of tank commanders. Its weak institutional position in the Soviet defence hierarchy partially explains why the Soviet Union opted early for missile delivery of tactical nuclear weapons, whereas NATO has remained largely wedded to air delivery.

aircraft, are not viable solutions for Europe. In short, while air forces can be expected to pay the price to maintain a permissive air environment, defence ministries must eventually decide whether the effort is worth the cost.

Mission analysis

Should tactical air power overcome the formidable problems of air defence, it still remains expensive, even with cheaper aircraft. The question to be asked then is: how does air power compare with new weapons systems competing for its roles? Its major roles are: (1) air superiority; (2) close air support; (3) battlefield interdiction; and (4) supply interdiction.

While Western air forces have focused upon *air superiority* in general and on the air-to-air encounter in particular, air superiority is in essence a derived mission. Its value depends on the effectiveness of air power's other missions: if these missions become low value, then air superiority and a demanding deep-penetration capability against well-defended and sheltered aircraft has little relevance to the overall conflict. This, of course, is part of NATO's military problem today: its air forces are devoting major efforts towards air superiority at the very time when its ground forces are relatively weak.

Although air-force doctrine has given precedence to long-range interdiction, *support* of the ground forces with weapons that will hold up tank operations is the crucial function in Europe. Close air support itself is essentially a substitute for artillery. Air power has had three advantages over artillery: (1) superior mobility; (2) an ability to destroy point targets such as tanks, through direct fire accuracy or through sheer size of munitions dropped; and (3) an ability to support fast-moving tank columns before army artillery can be brought forward and organized.[98] The disadvantages have been high cost, an expensive and difficult command and control system (both ground-controller to pilot and inter-service co-ordination, partly because of divergent service philosophies), and, most important from a ground commander's viewpoint, slow response and limited availability.

For close air support the two fundamental questions are: will ground forces continue to require shoring up and can other weapons do this more cost-effectively?

This study contends that by restructuring and introducing six relatively cheap technologies, so as to implement a chequerboard defence, army forces can themselves do much to immobilize the tank and stop enemy mobility. This meets most of the need for close air support; but close air support is certainly needed in crises where the army has been surprised or thrown in disarray, foreclosing an organized chequerboard defence.

But while this contention may seem bold, one of the six technologies is 'smart' artillery. With terminal guidance, artillery no longer needs a surrogate. Terminally-guided projectiles launched from howitzers, large mortars and multiple rocket launchers, guided by artillery and mortar forward observers (of which there are 117 in the current mechanized divisions compared with only 11 forward air controllers), are inherently low-cost, have a high rate of fire, and present no command and control difficulty (there are co-ordination difficulties inherent in close air support). The virtually zero incremental cost of 'smart' artillery is derived from the fact that the artillery system has to be there anyway. The added 'smartness' essentially allows the same artillery to stop the tank as well as the infantry. As for munitions cost, terminally-guided rounds are cost-effective: they may cost ten times more, but they are more than 100 times more effective. Finally, artillery's high rate of fire eliminates the value of air power's unique mobility advantage. Each forward observer can guide in a round every 10–15 seconds; he simply calls for a 10–15 second sequential stacking of coded projectiles aimed into the target's laser basket.[99] Theoretically, a forward division could, as has already been said, destroy all the armoured vehicles in an enemy division in a minute. The problem is then acquiring sufficient targets, not the ability to destroy them. With such a surfeit of destructive capacity, close air support adds little, for it too is controlled by forward observers and requires similar weather conditions.[100] Finally, while aircraft

---

[98] This last role has been eclipsed in British, French and American military thought, yet the original insight into the usefulness of tactical air power was German, with an innovating integration of tactical aviation and ground forces. According to this concept, tactical air power's prime role was in support of penetration tactics. In the scheme outlined in this Paper, this emphasis would be resurrected.

[99] Casualties among artillery forward observers have always been high (but lower than that of infantry-platoon leaders) because of their location in the frontlines. Laser targeting should not materially increase this rate and laser ranging for tanks and forward observers has already been introduced. Several new systems (e.g. *Hellfire*) also propose co-operative ground designation for air-delivered ordnance. Guidance lasers need to be held on the target only 5–10 seconds, just as for the *TOW* ATGM. The difference is that the laser beam is considerably more difficult to track back than is an ATGW, and protective cover can be very easily maintained by using a periscope.

[100] Weather is an inhibiting factor for laser and electro-optical guided weapons. In the worst winter months the air-delivered *Maverick* is limited to about 20 per cent of *daylight* hours. Laser-guided artillery is somewhat less constrained, because ceiling is less restrictive for artillery than ground visibility is for air delivery. Even more advantageous for laser-guided artillery, however, is that ceiling and designation time can be very low in the defence. With new survey techniques providing very accurate knowledge of firing and target

are not very useful in defending against the enemy's actual assault upon friendly positions because of battlefield confusion, smoke and the dangers from artillery (only friendly fire can be stopped, at the expense of its support), mortars with infra-red and radiometric seekers are all-weather, have a high rate of fire and are not tied to forward observers. In short, terminal guidance, by correcting the artillery's weaknesses, removes the basic rationale underlying much of the close air-support mission.[101]

Interdiction has the function of isolating forward forces from their reinforcements and supplies. Historically, interdiction has yielded mixed results. Supply interdiction has probably never been effective, but battlefield interdiction was quite effective for American and British forces in World War II. Its aim then was to disrupt transport communications behind the immediate battlefield in order to prevent the movement of reserves.

The *battlefield interdiction* mission can be accomplished in three ways: (1) inducing night-time confusion by sowing the road network with air- or artillery-delivered anti-tank and anti-personnel minelets; (2) attrition; and (3) blocking the road network. Because of the difficulty of destroying armoured vehicles in forested assembly areas and the short time available for attacking an armoured force in the interval between leaving its assembly areas and commitment to battle (plus the ready availability of replacements in penetration-style tactics), blocking the road network is the more effective means of battlefield interdiction. If the road system cannot be used, not only would logistic requirements be much greater, but the attacker would have difficulty in moving and co-ordinating his forces. Tactical air power with guided bombs of course accomplish this task, but so too can long-range artillery if stay-behind parties (e.g. overrun populations) can unobtrusively guide in 'smart' artillery. Even more significant is the potential of stand-off weapons, like ground-based DME and glide bombs. If these systems develop as their potential indicates, they will become the leading contenders for the battlefield interdiction role. If they do not develop, and tactical air power is unable to overcome air defence, the battlefield interdiction mission will go into limbo,

locations and with computers for rapid calculation of meteorological, gun wear, and ammunition-lot data, artillery CEP can be greatly reduced, allowing ballistic firing into a small laser envelope of only minimum terminal guidance.

[101] This conclusion needs underscoring, if only because so much effort is being expended upon co-operative systems whereby an air-launched stand-off projectile is handed to a ground controller for terminal guidance (by laser). The advantage of the co-operative system is that the pilot need not actually see the target, thus permitting greater stand-off range and reducing aircraft losses. This system has all the disadvantages of the forward observer plus the high costs and co-ordination difficulties of air delivery.

except for guerrilla-guided artillery fire and the night-time sowing of minelets.

The *supply interdiction* mission is already in limbo. Deep penetration into enemy air space requires sophisticated aircraft; without nuclear weapons they have little prospect of success. Aircraft cannot afford to loiter and destroy enemy trucks in an active air defence environment. Nor can the dense European transport grid be easily blocked. Moreover, if too many resources are committed to the deep interdiction mission (including air-base attack), as NATO has done by complicating the design of its aircraft with demanding penetration capabilities (high speed, large payload and long range, and sophisticated penetration avionics), resources are diverted from more central tasks, such as holding the attacker. And without a truly strong defence, NATO's expensively obtained supply-interdiction capability can always be offset with larger (and inherently cheaper) pre-positioned forward stocks. If supply interdiction is to be practicable, its solution requires cheap, stand-off weapons and target monitoring.

How should air power be constituted?

Restructuring ground forces and introducing new technology changes the role of tactical air power. Once deployed and re-equipped, army forces no longer need the same degree of air support; air defence is foreclosing the opportunities for interdiction by air, and there are in any case alternative ways of carrying it out. Tactical air power does, however, retain two critical roles: close air support and the derived mission of local air superiority. Even though manned aircraft may be more costly than competitive systems for normal defensive operations, tactical air power remains irreplaceable when friendly ground forces are disorganized or have been penetrated.

Close air support is therefore the only system available to blunt the enemy attack and gain time for the ground forces to sort themselves out and organize a coherent defence. It is also probable that during this decisive period the cohesion of the ground air-defence system is likely to be broken, and local air superiority will be required for protecting friendly close air-support aircraft, as well as for thwarting the attacker's ground-support aircraft and helicopter-borne infantry from further disrupting the defence.

Ground forces are most vulnerable to air power while moving: attacking forces therefore have a particular need for air defence and air cover. The two, however, have proved difficult to integrate. As long as the attacking forces are moving within their own lines, ground air-defence is probably sufficient for protection, though, of course, air cover can give added assurance by forward interception. The major role of air cover is protecting the counter-attack after a successful penetration into enemy-held territory and ground air-defence becomes inadequate.

The demands for close air support in the attack also vary with location. During the pre-attack phase close air-support demands are minimal. During the actual attack on prepared defensive positions, demands for close air support will be less than formerly, and will be focused upon hindering enemy movement and shifting dispositions. The demand for close air support peaks after the breakthrough, when the enemy is forced into a battle of manoeuvre while simultaneously losing air-defence cohesion.

For these two close air-support and local air-superiority missions, the Western practice of designing multi-purpose aircraft is no longer suitable. The requirement now is for two specialized, cheap aircraft – the lightweight air superiority 'dog-fighter' and the sturdy fighter-bomber for close air support. For air-superiority fighters the choice is between numbers and sophistication. Interceptors firing expensive long-range missiles may be appropriate – for the air defence of naval task forces for example – but not for battlefield air superiority. Sophisticated aircraft like the F-15 may be the best fighter in a one-to-one encounter and for an unconstrained budget, but they are less manoeuvrable than some of their light-weight competitors, too expensive, and thus foreclose the numbers and sortie rate necessary for local air superiority. Much of the sophisticated fighter's avionics are unnecessary and may actually reduce its overall effectiveness, given sortie-reducing maintenance demands and the vulnerability of its necessarily more demanding bases. The characteristic of close air support is that it is only feasible when the pilot can visually acquire and attack his target In countering breakthrough and penetration tactics, visual observation is particularly critical because of the interspersion of one's own and enemy forces, and the co-ordination difficulties inherent with a forward air-controller system attached to, and liable to collapse with, the army. Obviously, if close air support is only possible during daylight its supporting air cover does not need an all-weather attack capability.

Secondary considerations favouring the light-weight fighter are its better performance against helicopters and its greater capability to double as a ground-support aircraft. Enemy helicopters are not normally a threat to ground forces; however, in the confusion following a breakthrough they are likely to escape ground air-defences and can seriously aggravate the chaos in the defender's rear. Helicopters thus become a major interceptor target during a peak load period. The interceptor's problem with the helicopter is target acquisition; while look-down radars are perhaps necessary, they are often rendered ineffective

by ground clutter, and visual observation is thus crucial. The sophisticated fighter has no inherent visual advantage; lighter fighters in greater number are therefore preferable. The same reasoning applies to close air support; with terminal guidance reducing the value of large payloads, the sophisticated interceptor's greater payload is no longer inherently advantageous, and greater numbers of cheaper aircraft with a higher sortie rate are preferable.

The characteristics for good close air support are simplicity, survivability and lethality from the air force point of view, and responsiveness from the army's. The first three characteristics obviously favour fixed- rather than rotary-wing aircraft. The helicopter gained its popularity in the United States Army because the United States Air Force was unresponsive to army needs, and the lethality (payload and accuracy) of conventional ground-support weapons was neglected until the Vietnam conflict. Accuracy has now been rectified, but the responsiveness (or, as the army interprets it, dependability) issue remains a justification of an expensive attack-helicopter programme. Because of its limited range, speed and landing requirements, the helicopter is likely to be dispersed across the front near the forward forces. This dispersal and the consequence that pilots become familiar with the local situation ensures responsiveness to local operational needs (but requires local logistic support). Air force doctrine, on the other hand, argues that the centralization of aircraft and of their control is most economical and efficient; fighter aircraft are also, of course, more amenable to central control.

In infantry warfare, where units are more or less equally committed across the front, the army's wish for air support to be responsive to local needs has considerable validity. In armoured warfare and tactical nuclear warfighting the important criterion is, however, concentration at the critical breakthrough areas, and thus much of the justification claimed for the helicopter evaporates. Expensive aircraft should not be frittered away on secondary sectors; these assets should be centralized.

Finally – and perhaps decisively – the helicopter can only survive (even with less vulnerable rotor casings) by using ground-hugging tactics behind a definable battle line. But in breakthrough operations such a line may not exist (or at least may not be easily seen in the confusion of battle). The anti-tank helicopter is thus likely to be most useful only when it is least needed.

But though tactical air power should be centralized it should not be used on a continuing basis, since, even with cheaper aircraft, it is not competitive enough for the function. Close air support is likely only to be needed in strength at crisis points; at other times it may be redundant. This has two major implications. During crises tactical air power must be capable of very high sortie rates – which is easier for simple, rugged aircraft operating from forward bases. During non-crisis periods close air support will be little required and air-cover demands will be minimal. The vulnerable maintenance bases can therefore be located well to the rear (e.g. outside the guidelines area for Mutual Force Reductions in, say, the United Kingdom). Aircraft can then be largely based in safer areas during non-critical periods and deployed forward as crises develop.

Because air power is capital-intensive, changes in its role have major budgetary implications. Restructuring and infusing new technologies into NATO's ground forces changes their capabilities and their requirement for tactical air support. The penetration/ counter-penetration battlefield requirement is unsuitable for sophisticated aircraft like the F-15, MRCA and associated command and control aircraft such as the Airborne Warning And Control System (AWACS). Simpler aircraft, besides being more effective in support of ground forces, also imply major savings from changes in basing, maintenance, and pilot training (which governs peacetime operation and maintenance costs). How much this saves is difficult to calculate, partly because any major change requires a complete reassessment of how NATO generates its tactical air power. Comparisons of Swedish and Israeli air forces with those of NATO suggest that NATO's air forces are not well organized and that significant increases in aircraft per man and sorties per aircraft are possible.

These factors indicate that, even if NATO's present number of aircraft were retained, larger numbers of more effective aircraft could be concentrated upon the penetration battle, while procurement costs and manning could be readily cut in half. This translates into a 10 per cent manning and 20 per cent budget reduction for ground and air forces. The complication is that NATO's present inventory must be replaced by a new family of cheap air-combat fighters and rugged close air-support aircraft.[102]

Partially offsetting such savings is the need to provide more missile battalions to replace the nuclear-mission aircraft and the costs of the new stand-off interdiction missiles, though each of these can also be justified on other grounds – the first as rationalizing theatre nuclear forces and the second as opening up new capabilities. First generation interdiction missiles are expensive, but their costs should drop sharply and they are already cheaper than manned aircraft for heavily defended deep-penetration targets. On a peacetime costing basis, they should be particularly cheap. They require a large initial investment; but their manning, operations and maintenance costs should be low. Their manning and peacetime training costs can be partly met by converting existing nuclear missile battalions into dual-capable units. Hence a guess might be that, as far as the central front is concerned, a minimum 10 per cent manpower and budget cut would result from changes in the roles of tactical air power, while significantly improving close air support and acquiring new interdiction capabilities.

VII. CONCLUSIONS

A prime objective of military analysis in recent years has been cost-effectiveness. Such analysis seeks incremental changes from accepted procedures, but by its nature it has little impact upon the general military balance and upon questions of political significance. In this Paper fundamental changes have been proposed for both doctrine and the application of new technology, so as to provide a quantum jump in military capabilities, and significantly affect the military balance and remove destabilizing asymmetries. Conventional parity (and a fortiori defensive superiority) revises the premises underlying much of the analyses of Atlantic and East–West relations. A true conventional balance, primarily European, would simplify and even solve many of the Alliance's most pressing politico-military difficulties. A new set of political problems would no doubt emerge, but at least Europe would be negotiating from a stronger position and the United States

would have reduced the saliency of her European burden and commitment.

Various schools of thought have sought to explain NATO's conventional deficiency and find a solution. Among the more popular ideas has been the one – particularly appealing to Americans – of using technology to bridge the gap. Tactical nuclear weapons, for instance, have seemed to offer an attractive way to substitute firepower for manpower. The military have certainly been drawn towards the notion of offsetting inferiority in numbers of combat forces by superiority in equipment. Then there is the school that holds that modern anti-tank weapons give, perhaps for the first time, the possibility of a new role for the citizen soldier in the defence, to

---

[102] Impending decisions on aircraft replacement are therefore particularly crucial at this time; aircraft designed for dated missions should not be bought, as they would lock NATO into an undesirable tactical air posture for the next decade.

enable him to replace some of NATO's expensive standing forces.

Perhaps the most influential school – at least to outward appearances – has been that of the economists. The dominant characteristics of this economic analysis are its use of resource inputs to measure the problem and the prescriptions of economic theory to solve them. This approach has led to a two-part conclusion: that there is a conventional balance in Europe – or there nearly is – and, since at worst the shortfall is only marginal, it can be explained by the Alliance's failure to make the best collective use of its military expenditures through economic rationalization and specialization.

None of the above approaches has commanded universal acceptance because of their apparent inconsistencies. The tactical nuclear and technological superiority schools seek to offset a conventional deficiency by spending more, when NATO is outspending the Warsaw Pact already. The economic school has recognized the paradox of spending more and obtaining less, but ends up either denying the existence of the problem or explaining NATO's deficiencies as manifestations of incongruities with economic theory. The school which would rely more on civilian soldiers seeks a radical departure from present military concepts, but its solution is militarily suspect and strategically inadequate for an area as vital as Germany.

All of these approaches skirt NATO's doctrinal and organizational precepts. None comes to grips with why NATO requires *wartime* divisional slices of over 40,000 while the Warsaw Pact manages with a slice only half as large but with the same combat strength – and that for the more demanding role of the offensive, with a greater reliance on tanks and artillery. None highlights the fact that NATO's concept (of relatively few but high-quality sustainable combat forces) and the Warsaw Pact's concept (of larger numbers of smaller less sustainable forces equipped with cheap rugged equipment) represent quite different warfighting philosophies.

Restructuring seeks to come to grips with these doctrinal and organizational realities. It does not deny the importance of tactical nuclear weapons, the usefulness of economic tools, nor that militia has a part to play. But whereas each of these approaches takes us only so far, restructuring seeks to reap the maximum advantage from the analysis of how force can and ought to be employed within present and likely future technological constraints.

Restructuring and the potential of new technology lead to the conclusion that a cross-over may soon occur in the age-old struggle for ascendancy between the offence and the defence. The offence, through the mechanisms of the fighter-bomber and the tank, has been dominant for the last forty years. New technology now entering service, accompanied by necessary doctrinal and organizational changes, can challenge this dominance and give it, as far as conventional warfare on land is concerned, to the defence instead.

This Paper has therefore argued that:

(1) by changing concepts and organizations NATO (including France) can obtain a true capacity for mobile defence and a conventional balance with the Warsaw Pact;

(2) with the additional use of current cheap technology NATO can implement a genuine forward defence and obtain defensive superiority in both conventional and tactical nuclear warfare;

(3) with current and turn-of-the-decade technology NATO can enhance this defensive superiority, yet reduce ground and air manpower on the central front by as much as 15 per cent, as evidenced by technologically induced savings in artillery and tactical air power. Technological infusions in other branches should lead to further reductions.

2 New Weapons Technologies: Debate and Directions

RICHARD BURT

INTRODUCTION

Technology is recognized as an important but elusive factor in warfare. In single battles and campaigns the technological prowess of opposing forces has been a crucial determinant of victory or defeat, while in the longer term technology has been seen by military historians to fit cyclical patterns of offensive and defensive ascendency. By shaping the tactics and strategy best suited to the exercise of military force, technology has also profoundly affected international politics: by setting limits on armed forces' capabilities to seize or defend territory and to undertake other operations, it has moulded political intentions and expectations.

For modern industrialized states in the twentieth century, however, military technology has played an even more vital role in the use and organization of armed forces; as in other spheres, constant and rapid exploitation and substitution of new technologies have yielded order-of-magnitude improvements in capabilities. The most striking development in the post-war era, of course, has been the development and refinement of fission and fusion weapons, whose firepower has shaken traditional assumptions about warfare and placed a new emphasis on the mechanisms of deterrence. Beneath the nuclear threshold, however, developments in conventional weapons have been almost equally dramatic.

The process of rapid technological substitution has not only led to enhanced capabilities but has also gradually turned Western armed forces into technocracies where a declining ratio of combat to support personnel has meant that, though firepower has increased, a decreasing number of individuals are actually involved in combat. This trend holds for the forces of all nations, but it is most apparent in Western forces, particularly those of the United States.[1] The primary explanation is that Western nations have traditionally had a comparative advantage in technology – and, since these nations possess capital, scientific expertise and a strong industrial base, the exploitation of new technologies is viewed by many as the primary means by which they can compensate for weaknesses or deficiencies in other areas.[2] As General George Brown stated succinctly: 'The United States has never attempted to match the Soviet Union in either ground force personnel or material, relying instead on technology'.[3]

Despite this tendency in western military thought, several critics have begun to question the emphasis given to advanced technology in defence planning. Arguments have focused especially on the way technology has been used in recent years. It is maintained that in some areas of weapons design improving performance characteristics has become an end in itself, and

[1] See William D. White's discussion of 'Technological Substitution in the Armed Forces', *US Tactical Air Power* (Washington DC: The Brookings Institution, 1974), chapter 2.
[2] Perhaps the most strongly worded official statement of this view comes from Dr Malcolm Currie, Director of Defense Research and Engineering: 'In this increasingly competitive, often hostile and rapidly changing world Americans seem to have only one real choice. Clearly our national well-being cannot be based on unlimited raw materials or on unlimited manpower and cheap labour. Rather it must be based on our ability to multiply and enhance the limited natural and human resources we do have. Technology thus appears to offer us our place in the sun – the means to insure our security and economic vitality.' *Program of Research, Development Test and Evaluation, FY 1976* (Statement submitted to the US Senate Armed Services Committee), 7 March 1975, p. 1–2.
[3] *United States Military Posture for FY 1976*, statement by Chairman of the Joint Chiefs of Staff, Gen. George Brown USAF, p. 70.

that design parameters have become detached from battlefield requirements: for example, combat aircraft used in Vietnam were faster than their predecessors, but this hindered rather than helped target acquisition and munitions delivery. Moreover, technological substitution is not cheap: it is now a truism that for any major weapon a follow-on system of the same type will be significantly more expensive. Does it make sense, then, to spend in many cases twice or three times as much for a replacement promising only marginal performance advantages over an existing system, particularly when this cost rules out one-for-one replacement? There is also the cost of manpower. While reduction of personnel costs has been a major argument in favour of technological substitution, in some cases the result has been just the opposite. Studies have shown that over half the ten-year 'life cycle costs' of modern weapons are for operations and maintenance.[4]

These arguments come at an opportune time, for a new generation of technology has stimulated a new debate on the role and implications of technology for warfare.[5] Developments in data storage and retrieval, sensor technologies, integrated microcircuits, materials science, new propulsion technologies and laser applications all offer a new range of possibilities. By holding out the promise of radical new approaches to the performance of existing missions, substantial cost advantages may appear. Constructing systems along modular lines with expendable components may make greater reliability and reduced operations and maintenance costs feasible. But

most intriguing is Dr Malcolm Currie's idea that 'a remarkable series of technical developments have brought us to the threshold of what will become a true revolution in conventional warfare.'[6]

However, the mere existence of new technologies, and even new weapons, is no guarantee of the dramatic change that Currie suggests. Looking back, it is clear that certain technologies – the machine gun and the submarine in World War I, the tank and the strike aircraft in World War II – had a crucial impact on the style and pace of military operations, and more specifically on the nature of the duel between offense and defence. But, looking ahead, the impact of the new technologies will ultimately depend on how soon and in what form they are used; this in turn will depend on the resolution of a number of complex technical, organizational and political questions that must be examined in more detail than is attempted here. This study, then, does not endeavour to supply firm answers to these questions, but it does attempt to provide a general framework for judging their relative importance. In order to do this, major technological developments are outlined in Part I, and the important arguments that have been used to predict their impact are set out in Part II. The two subsequent sections will examine different elements of these arguments: Part III covering operational and organizational implications of the new technologies, and Part IV their possible impact on political arrangements, arms control efforts and arms transfer policies.

I. THE NEW WEAPON DEVELOPMENTS

In times of relative peace revolutions in military technology happen slowly, and the belief that a breakthrough in technology is now being made belies the fact that conventional weapons have been continually improved since World War II. Many of the techniques that were 'proved' and the concepts 'learned' in the 1973 Middle East war existed or were formulated over a decade

earlier. But the shock of actual conflict – in the Middle East, Indochina and elsewhere – has awoken a new interest in exploiting weapons technologies, many of which are not necessarily novel in themselves. Thus, while recent experience has spurred fresh development efforts, experience suggests that the introduction of new equipment (and doctrine) will, in most cases, proceed slowly and incrementally.

[4] 'Life cycle' accounting is a comprehensive calculation of the total costs associated with the development, procurement and operation of a system over its full period of service.
[5] This analysis is concerned with the technologies now being adapted, or likely to be exploited, for military purposes in the coming decade. It will not therefore dis-

cuss emerging techniques that could have a major impact further in the future, such as high-energy lasers and large-scale weather and climate modification.
[6] Quoted in Phil Stanford, 'The Automated Battlefield', *New York Times Magazine*, 23 February 1975, VI, p. 12.

At the same time, it is clear that weapons technology is evolving rapidly and that the diversity of this change is impressive. In 1975 the United States alone funded over thirty new conventional weapon families at the advanced stage of development; and in the area of anti-tank guided weapons (ATGW) alone more than eighteen new systems are under development by NATO members. These developments not only promise greater firepower, mobility and protection – traditional goals of weapon design – they could also introduce more control and flexibility into combat operations, by substantially upgrading surveillance and target destruction at increasing ranges, while reducing collateral damage to civilian structures and population. The most significant developments seem to be taking place in the following areas: precision guidance; remote guidance and control; munitions improvements; target identification and acquisition; command, control and communications; and electronic warfare.

Precision Guidance
Perhaps the most striking of the developments is the increased accuracy obtainable from new and refined guidance techniques. The term 'Precision-guided munitions' (PGM) is now used to describe a growing class of bombs, missiles and artillery projectiles with single-shot kill probabilities from ten to a hundred times greater than unguided munitions.[7] As Table 1 shows, this increase in accuracy is made possible by numerous guidance technologies that can reduce the circular error probable (CEP) of delivery vehicles to 20 metres or less.[8]

Several writers have offered *operational* definitions for PGM that identify these weapons in terms of their probability of hitting desired targets (50 per cent or more). While these descriptions have drawbacks, they do convey the essential element of precision guidance: that incorporating sensing technologies into a wide range of munitions promises almost 'one shot, one kill'.

For certain roles, PGM have existed for over 20 years: accurate surface-to-air missiles, like the

Table 1: Estimated Accuracy of Precision Guidance Technologies, 1975–85

Guidance Technology	Accuracy (metres)
Laser designation	10
Electro-optical	10
Infra-red seeker	10
Radar homing	50
Radar Area Correlation	50
Distance Measuring Equipment	50
Microwave radiometer	20
Satellite position fixing	10

NOTE: Accuracy figures are drawn from estimates released by the US Department of Defense, the Raytheon Corporation and the Lockheed Missiles and Space Co., Inc. In some applications (such as distance measuring) accuracy is inversely related to range, which has not been specified in public estimates.

US Navy's radar-guided *Terrier*, and the Soviet infra-red-homing SA-2 were deployed in the early 1950s, as was the French SS-10 wire-guided anti-tank missile, and the US Air Force's infra-red-homing *Falcon* air-to-air missile was deployed in 1955. The most important developments of recent years have been to improve the performance of guided systems in these roles and to harness sensor technologies to the air-to-surface role, where target acquisition and tracking is a more serious problem. For both airborne and surface (land or sea) targets, efforts have been profitably directed towards improving the accuracy of longer-range systems and building precision guidance into smaller (often man-portable) battlefield weapons. Table 2 shows that missiles employing advanced guidance techniques now vary widely and are deployed or are under development by several nations. Guidance techniques can, however, be grouped into three categories: seeker guidance, precision positioning and correlation guidance.

Seeker Guidance
PGM using these systems, which home on fixed or moving targets,[9] include glide bombs homing on energy reflected from a target illuminated by a laser beam, missiles with infra-red seekers that

---

[7] Precision-guided bombs have been nicknamed 'smart' bombs, to distinguish them from their less intelligent unguided predecessors.
[8] The CEP is the radius of a circle around a point target within which half the weapons can be expected to strike.

[9] More precisely, sensors enable PGM to home on the target's energy 'signature'. The signature can be *natural* (like the heat from an aircraft engine) or *induced* (like the reflection given off by a target designated by a laser beam), or it may be determined by the *contrast* of target and background (like a dark aircraft flying against a clear sky).

Table 2: Extent of Indigenous Development of Advanced Missiles, 1975[a]
(D = under development; s = in service)

Country	Surface-to-surface ballistic	Long-range cruise	Air-to surface	Anti-ship	Surface-to-air	Portable anti-tank	Air-to-air	Anti-submarine
USA	D	D	s	s	s	s	s	s
USSR	D	s	s[b]	s	s	s	s	s
France			s	s	s	s	s	s
Britain			s	D	s	s	s	
W. Germany			D	D	s	s	D	
Israel			D[c]	s			s	
Italy				s	s	s	D	
Sweden				s	D	s		
Australia								s
Japan				D	D	s	s	
Norway				s				
Canada					s			
S. Africa							s	

[a] For the purposes of this table, 'advanced' covers systems using seeker guidance, precision positioning, correlation guidance and stellar inertial guidance (but not other forms of inertial guidance). Not shown in the table are advanced systems which the countries listed have in service, but which were obtained from abroad and not developed indigenously.
[b] Includes only long-range, command-guidance missiles, though the AS-5 *Kelt* may have an active radar-seeker.
[c] A TV-guided missile, the *Rafael*, is reported to be under development.

home on the hot metal or emission of a target's exhaust, systems that employ active and passive radar, and (in the future) sensors to track and engage microwave-emitting targets.[10]

Although seeker guidance has been used primarily for missiles, terminal-guidance capabilities are now being wedded to several types of projectiles. Under development for the US Army's 155mm and 8in. howitzers are cannon-launched guided projectiles which will home on targets illuminated by laser designators operated by forward observers (accuracies of 1–3ft have been obtained in tests),[11] and a laser-guided mortar round is being tested by West Germany. Laser and infra-red applications are also under development for low-altitude anti-aircraft and naval guns. Warheads composed of terminally-guided submunitions are being designed for delivery by tactical ballistic missiles. When released from a warhead dispenser, these projectiles would home on targets at distances of over a mile.

Precision Positioning

This uses signals from synchronized transmitters or beacons to correct the accumulated errors of inertial guidance systems aboard delivery vehicles. At present the accuracy of beacon technologies, such as the Long Range Navigation (LORAN) system, only permit mid-course guidance. However, position-fixing devices, when coupled with terminal guidance sensors which would take over after the mid-course guidance phase, promise to give long-range missile systems accuracies similar to battlefield PGM. But newer positioning systems, such as distance-measuring equipment (DME), may alone yield terminal guidance accuracies. An advanced version of the *Lance* ballistic missile under consideration by the US Army, for example, would use synchronized signals received from airborne beacons to guide the warhead towards a target area.[12]

[10] Unless otherwise stated, descriptions and characteristics of systems mentioned in this section have been taken from: 'World Missile Survey', *Flight International*, 18 May, 1975; aircraft and missile tables in *Aviation Week and Space Technology*, 17 March, 1975; *Jane's Weapon Systems, 1974–75* (London: Jane's Yearbooks, 1974); *The Military Balance 1975–76* (London: IISS, 1975); *Weapons Technology: A Survey of Current Developments in Weapons Systems*, (London: Brassey's, 1975) and 'Gallery of USAF Weapons', *Air Force Magazine*, May 1975. A dated, but still useful, introduction to missile guidance techniques can be found in M. Allward and J. W. R. Taylor, *Rockets and Missiles*, (London: Ian Allen, 1958). Current and emerging guidance technologies are surveyed in more detail by James Digby, *Precision-Guided Weapons*, Adelphi Paper No. 118 (London: IISS, 1975), appendices I and II, pp. 14–24.
[11] *International Defence Review*, April 1975, p. 262.
[12] See 'Lance Tactical Warhead Ordered', *Aviation Week and Space Technology*, 6 October 1975, pp. 14–17.

The most important development in this area, however, would be the use of satellites to supply course-correcting data. Using 24 satellites in stationary orbit, the American Global Positioning System (GPS) will reportedly permit a delivery vehicle's position to be determined within about 20ft in latitude, longitude and altitude at intercontinental ranges.

Correlation Guidance
These guidance systems are essentially 'map-matching' devices, particularly useful for long-range guidance against known targets that either cannot be easily designated or do not have a strong enough signature for homing systems. Using optical, radar, infra-red or microwave sensors, correlation systems compare sensed pictures of a target area with stored reference pictures to generate course corrections.[13] A system using radar for mid-course guidance is under development for the *Lance*, and a television correlation device using aerial reference photos has been tested for use on the *Pershing* tactical ballistic missile.[14] The long-range cruise missiles under development in the United States could employ a so-called terrain contour matching (TERCOM) approach, which would use a radar altimeter for mid-course guidance and a microwave radiometer to measure terrain reflectivity for the terminal phase. It is estimated that this combination would give accuracies of 30 metres or less at intercontinental ranges. Like the GPS, TERCOM-type systems are at an early stage of development and are unlikely to be available for deployment before the mid 1980s.[15]

Remote Guidance and Control
Like PGM, remotely-piloted vehicles (RPV) have existed for many years, but only recently have developments in the design of electronic data links permitted the execution of complex missions. Used for reconnaissance by Israel in the Middle East and by the United States during Indochina conflict, RPV appear to hold the greatest promise for target designation or strike and for electronic warfare. Recoverable RPV can now be used as support jammers and chaff dispensers to degrade defensive radar networks, and low-cost expendable drones (without remote piloting) will soon be available to saturate air defences.

Strike RPV, coupled with stand-off PGM, could be used against a variety of targets. A test-bed for such a system is the American *Firebee* supersonic RPV, directed by a television camera in its nose, which has successfully launched the *Maverick* PGM against moving tanks.[16] Other advanced RPV under development in the United States include the long-range, high-altitude *Compass Cope* reconnaissance vehicle. With a planned flight endurance of over 30 hours, this could become a primary means of high altitude surveillance and electronics monitoring in the 1980s.

One of the most interesting options now under study is the 'mini RPV', a propellor-driven battlefield reconnaissance craft. A system now under consideration would weigh only 45 lb and could be air-launched and controlled by pilots to acquire targets for attack. It could also be launched and recovered by ground units and used for reconnaissance or to designate targets for missiles or laser-guided artillery.

Improved Munitions
Though overshadowed by increased accuracy and improvements in remote guidance, continued refinements in conventional munitions offer enhanced destruction capability combined with reduction of unwanted blast effects.[17] One important development has been to give conventional munitions greater reliability by certifying

[13] For a more detailed explanation of correlation guidance methods, see Kosta Tsipis, 'The Long Range Cruise Missile', *Bulletin of the Atomic Scientists*, April 1975, pp. 15–26.
[14] 'Guidance Device Set for Pershing Tests', *Aviation Week and Space Technology*, 12 May 1975, pp. 45–48.
[15] See Edgar Ulsamer, 'Technological Initiative: A Priceless Asset', *Air Force Magazine*, June 1975, p. 31.

[16] See Irwin Stabler, 'US Remotely Piloted Vehicle Programs', *International Defence Review*, April 1974, pp. 180–182. This article includes a comprehensive discussion of RPV avionics and fire-control systems under development in the United States. For a survey of different national RPV development programmes, see 'Drones and Remotely Piloted Vehicles', *World Armaments and Disarmament*: *SIPRI Yearbook 1975*, (Stockholm: Almqvist & Wiksell; Cambridge, Mass., and London: MIT Press, 1975), Chapter 12.
[17] This section draws heavily on C. I. Hudson's excellent discussion of conventional weapons effects, 'Improvements Possible with New Conventional Munitions Technologies', in Johan Holst and Uwe Nerlich (eds), *Beyond Deterrence*: *New Aims, New Arms* (New York: Crane, Russak, forthcoming 1976).

production techniques for warheads, propellants and fusing systems. Increased delivery accuracy, the profusion of types of munition for point and area targets and higher reliability have brought about a marked improvement in 'weapon-tailoring' capabilities (the ability to match a specific target with the most efficient munition to destroy it). For point targets like command posts, bridge piers and individual tanks, kinetic-energy or explosive penetrators and armour-piercing munitions can be selected according to the desired effect. 'Hard structure' weapons, with accelerated terminal velocity projectiles to penetrate several feet of concrete, can be used against hardened bunkers. 'Earth penetrator' weapons, which burrow several feet under hard surfaces, are effective against roads and air strips.

For targets that require large area coverage, fuel-air explosives can spread blast energy more uniformly than conventional TNT. Conventional high-explosive (HE) munitions can, however, be designed to maximize either blast or fragmentation damage, or to provide a particular mix of both, depending on the target. In more sophisticated weapons-tailoring techniques, combined effects can be used. Combined blast and incendiary munitions might be employed against fuel storage depots, for example, with the blast effect exposing volatile substances to the air so that they can be ignited by the incendiary.

For more efficient coverage of area targets, munitions can also be packaged as clusters and dispersed in large numbers. The potential of terminally-guided sub-munitions has been noted, but even with moderate 30–50m delivery accuracies, unguided sub-munitions yield more efficient target coverage with less collateral damage than larger unitary bombs. Another cluster technology of growing promise is mine implacement. Cannisters that can carry as many as 80 light-weight 'minelets' can be dropped from aircraft or delivered by artillery. A new generation of West German and American mines, designed to respond optimally to different targets, can also be activated or switched off by remote control: the *Gator* mine comes in indistinguishable armour and anti-personnel versions, while the *Grasshopper*, for use against logistics vehicles, buries itself upon impact and only tosses a warhead above ground when a target comes within its lethal radius. Barrier weapons like mines also provide the means to use other area weapons more efficiently, by impeding or stopping advancing units and thus improving targeting opportunities.

Target Identification and Acquisition

Detection, location and targeting are essential to highly accurate systems, and target acquisition functions are integral to many PGM systems. Anti-radiation missiles seek out emitting targets; infra-red-guided systems home on targets giving off heat; and with simpler systems, like infantry anti-tank guided weapons (ATGW) and man-portable surface-to-air missiles (SAM), the operator acquires and tracks targets visually. Direct observation can also be used for longer-range guided systems, like stand-off missiles or cannon-launched projectiles by equipping forward observers with man-portable laser designators. RPV could provide acquisition both for battlefield targets and for those far beyond enemy lines. Long-range and high-endurance RPV could be used to detect incoming missiles and aircraft, as a target designator for tactical missiles, or to provide beacon positioning data. In a maritime role, long-range RPV could monitor extensive areas of sea.

American Airborne Early Warning and Control System (AWACS) aircraft will carry long-range surveillance radars to provide 200-mile over-the-horizon coverage, including the detection of low-flying enemy aircraft. At shorter ranges, moving target indication (MTI) radar can detect ground forces massing for attack and imaging radars can locate ground targets. For naval forces, the US Navy's integrated fleet-defence system, *Aegis*, incorporates several advanced radar systems to acquire and track multiple aircraft and missile targets. Coupled with a computerized fire-control system, it can direct point and area defences of several vessels operating together.

On the battlefield, 'infyonics' – the application of sensor and communications technologies to small land units – has, until recently, been a little-exploited area.[18] Some impressive advances are now available, however. Electronic detection barriers (unattended ground sensors responding to noise, metal, seismic pressure, and even smell), radar-assisted rifles, hand-held radars, night

---

[18] See Brig. Gen. F. P. Henderson USMC (Ret.), 'The FMF: Alternative Future and How to Get There', *Marine Corps Gazette*, July 1971, pp. 18–48.

vision scopes and thermal imagers improve the ability of ground combat units to find and identify targets in bad weather or by night.

Command, Control and Communications

Several related improvements in information collection and transmission promise commanders a more comprehensive view of the theatre of operations, while advances in miniaturization will enable small units to receive and transmit urgent information. On the theatre level, the effort is being made to enable commanders to make strike decisions on the basis of real-time intelligence.[19] This has been facilitated by computer developments that allow rapid synthesis of reconnaissance information. Under NATO's *Seek Bus* programme, all tactical air elements will be linked in a network giving commanders rapid information on allied and enemy deployments, weapons and targets. In central Europe, a similar effort is under way to establish a single integrated air-defence command to control and allocate air defence resources.[20] A key system here will be AWACS, which will assume command-and-control functions where there are now gaps in air defence or where command facilities are lost. For ground forces, computerized battlefield command-and-control systems are being designed to give field commanders rapid access to information on enemy movements, aircraft and artillery availability, weather conditions and logistics support. At sea, the *Aegis* fleet-defence system will serve a similar function by integrating the defences of a naval task group and controlling the launch of several different weapons.

On the global level, a major effort is being made in the United States to upgrade the Worldwide Military Command and Control System (WWMCCS). This includes modernization of computers, deployment of an advanced Airborne Command Post, improvements to communications and greater interoperability between the communications of the separate services.

Important developments for individual weapons are under way in the design of data links to transmit complex commands, but which are small, cheap and reliable under combat conditions. Particularly important are spread-spectrum techniques, enabling data to be transmitted over changing frequency bands in order to defeat jamming.

Electronic Warfare

Efforts to control the electromagnetic spectrum for military purposes have been a feature of conflicts throughout this century. Only in the last decade, however, have the three basic tasks of electronic warfare – electronic intelligence, electronic countermeasures (ECM) and electronic counter-countermeasures (ECCM) – assumed central importance. During the invasion of Czechoslovakia in 1968 the movement of Warsaw Pact ground and air forces was effectively masked by massive barrage jamming of whole frequency bands; in the air war in Indochina the United States used a combination of point jamming (of exact radar frequencies) and deception techniques aboard aircraft to counter North Vietnamese radar-controlled anti-aircraft guns and SAM. It was during the 1973 Middle East conflict, however, that electronic warfare was shown to have a crucial role in modern combat. Both sides engaged in widespread monitoring of communications, as well as ECM and ECCM. 'Point' jamming was directed against air-defence radars whose frequencies were known, or could be detected, and 'barrage' jamming was successfully used against tanks to disturb communications. At sea and in the air, Israel also used chaff to screen ships and aircraft from enemy radars.[21]

The widespread use of electronic warfare in the 1973 Middle East conflict has spurred numerous developments, particularly in the United States and the Soviet Union. All new Soviet SAM can now vary their radar frequency, which makes point jamming more difficult, and the so-called 'frequency agility' of new radar-controlled SAM systems has led to the development of computer-directed jamming systems with greatly improved reaction times. Because both jamming and jamming counter-measures require considerable power, electronically-steer-

---

[19] Real-time command and control means the ability to transmit, receive and act on data as events take place.
[20] See Gen. John E. Ralph USAF, 'Tactical Air Systems and the New Technology', in G. Kemp, R. Pfaltzgraff, U. Ra'anan (eds) *The Other Arms Race* (Lexington, Mass.: Lexington Books 1975).

[21] Chaff, first used during World War II, consists of thin metal or plastic rods which are dispensed into the air and appear on radars as false targets. To work properly, the length of the rods must correspond to half the wavelength of the radar's frequency.

able directional antennae are being used to reduce the power waste associated with crude barrage jamming.

Aircraft and ships are now being fitted with new electronic warfare equipment, but new systems are being designed specifically for this mission. The US Navy now uses a modified A-6E aircraft for stand-off jamming missions, and the US Air Force is developing the EF-111A, a version of the swing-wing F-111, to provide deep-strike interdiction aircraft with escort jamming of air defences.

II. A NEW ERA OF WARFARE?

Taken together, the foregoing developments have led a growing number of observers to suggest that a new era in conventional warfare is on the horizon. Thinking on the subject is perhaps at too early a stage to allow a clearly definable school of thought to be identified but, if any central theme has emerged from recent analyses, it is that, theoretically, the technologies will favour defence over attack. In this section, the various arguments used to sustain this claim will be outlined.[22]

According to several writers, the most important aspect of the new weapons is the greater killing power available to small independent units: the new technologies make fixed bases and large and costly systems like tanks, strike aircraft and surface warships more vulnerable to detection and targeting by the new generation of ATGW, SAM and anti-shipping missiles, while enabling small units to make use of benefits inherent in defence, such as concealment. With the new technologies, a target that betrays its location is said to have a high probability of being destroyed, so that a greater premium is said to be attached to 'hiding, blending with the background, and remaining motionless'. 'This premium obviously is of greater benefit to a defender, who can play a more passive role, than an attacker. The latter is obliged to move aggressively forward, often over open terrain in unfamiliar territory. The opportunities for the defender to locate the attacker and bring him under PGM fire are inherently greater.'[23]

Significant tactical implications are seen to flow from these observations. Possibly the most important is the suggestion that both defensive and offensive forces, in the words of various writers, will become increasingly 'decentralized', 'molecular', and 'federated'. Defensive forces will operate in larger numbers of dispersed groups, because of the growing firepower of smaller weapons and the importance of exploiting the advantages of hiding. For the attacker, the greater vulnerability of larger systems, operating on or over the battlefield or resting in tank parks or on air strips beyond the battle area, is thought to necessitate dispersal tactics which will reduce the 'shock power' of invading armies and make the concentration of forces for the classic land battle 'breakthrough' a risky enterprise. Under these circumstances, the defender, deployed in a familiar environment, able to make good use of natural and constructed cover, and using a new array of weapons, is said to enjoy the advantage. One writer has elaborated a 'chequerboard' defence concept, where small, dispersed strongpoints would use man-portable

----

[22] Analyses of the politico-military impact of the new technologies are few, but the number is steadily growing. See particularly Digby, *op. cit.* in note 10; Donald Brennan (ed.) 'The Implications of Precision Weapons for American Strategic Interests', Hudson Institute (HI-2204) January 1975; Steven Canby, *The Alliance and Europe: Part IV, Military Doctrine and Technology*, Adelphi Papers No. 109 (London: IISS 1975); John T. Burke, 'Smart Weapons: A Coming Revolution in Tactics', *Army*, February 1973, and 'Precision Weaponry: The Changing Nature of Modern Warfare', *Army*, March 1974; Albert Wohlstetter, 'Threats and Promises of Peace: Europe and America in the New Era', *Orbis*, Vol. XVII, No. 4, Winter 1974; Peter A. Wilson, 'Battlefield Guided Weapons: The

Big Equaliser', *Proceedings of the Naval Institute*, February 1975; Frank E. Armbruster, 'The Effect on Defence Policy, Strategy and Tactics of the Introduction of Smart Bombs and Smart Missiles into the Arsenals of the World', Hudson Institute (HI-1799/3-DP) October 1974; Henderson *op. cit.* in note 19; Kemp, Pfaltzgraff and Ra'anan, *op. cit.* in note 21; Edward B. Atkeson, 'Precision Guided Munitions: Implications for Détente', *Parameters*, Vol. V, No. 2, 1976; Col. Stanley D. Fair, 'Precision Weaponry in the Defence of Europe', *NATO's Fifteen Nations*, August–September 1975; Horst Menderhausen, 'Inoffensive Deterrence', *California Arms Control and Foreign Policy Seminar*, May 1974.

[23] Edward B. Atkeson, *op. cit.*, p. 9.

weapons and remote sensors to impede heavy armoured thrusts.[24]

Some writers have predicted a similar development at sea. With large, easily-located surface vessels, such as aircraft carriers, becoming more vulnerable to small, manouverable, PGM-equipped vessels, it is suggested that the firepower of a major surface combatant might be achieved by several smaller vessels carrying anti-shipping and anti-submarine warfare (ASW) weapons operating in a spread-out formation.[25] It is also maintained that amphibious assaults against hostile fire could also become more difficult, as ground-support aircraft, helicopters and emphibious landing craft become less able to survive in the face of PGM fire.[26]

The tactical consequences of the new technologies thus appear to open up several exciting possibilities. They seem especially relevant to defence in Western Europe, particularly in the central region, where the problem has long been one of countering the Warsaw Pact's superior numbers of infantry, artillery and, most important, tanks. Improved command and control, target acquisition aids and air-delivered and ground-borne PGM are thought to be able to give NATO forces the ability to knock out substantial numbers of tanks in a short time.

Blunting an armoured attack in the central region is the chief advantage claimed for the new weapons, but others are also cited. Firstly, the increased kill probabilities of the new systems, especially SAM and anti-shipping missiles, seem well-suited for the defence of coastal areas, like the Baltic, and air bases in lightly-defended regions, such as northern Norway; in the south, man-portable ATGW appear ideal for use in the difficult terrain of the Anatolian Peninsula.

Secondly, higher accuracies and new conventional munitions hold out the prospect of substantially reducing collateral damage in an intense conventional war. It therefore has been suggested that the new weapons could change the attitudes of the central region's population towards preparing for such a conflict, thus for the first time enabling a 'defence in depth' approach to NATO strategy to be adopted. Thirdly, the availability of accurate man-portable weapons, easy to operate and maintain, would seem to make feasible another long-discussed idea: handing over a greater defensive function to militia-type reserves, reducing the demand for trained manpower and freeing standing forces for more complex tasks. Finally, the comparatively low costs of some of the new systems ($3,000 for a *TOW* ATGW, $10,000 for a *Stinger* SAM) suggests that abundant numbers of the new systems could still be procured in an era when NATO defence expenditure is unlikely to increase substantially in real terms.

Significant military implications for other states are also foreseen. By improving defensive capabilities, the new weaponry is thought to make weaker states better able to resist invasion from stronger neighbours and intervention by outside powers. The systems thus seem particularly relevant to states with special defensive problems, such as Yugoslavia, Sweden, Pakistan or Japan, which must coexist with far more powerful neighbours.[27] For nations like Israel, which have a high degree of technical competence, the new technologies might enable armed forces to use technological advantage to counter less sophisticated but numerically stronger adversaries. More intriguing, however, is the argument that less-developed states with

---

[24] Steven Canby, *op. cit.*, pp. 23–7. The use of chains of mutually supporting anti-tank units to counter armoured thrusts is not a new concept. In the 1930s J. F. C. Fuller elaborated the concept of the 'archipelago defence' to counter shock offensives, where small anti-tank units, deployed in depth, acted as 'islands' to canalize and then halt armoured penetrations. See *Armoured Warfare*, (London: Eyre & Spottiswoode, 1943) and Maj. L. Wayne Kleinstiver, 'The Archipelago Defense', *Infantry*, March–April 1974.

[25] Hubert Feigl, 'New Naval Technologies', in *Power at Sea, Part I: The New Environment*, Adelphi Paper No. 122 (London: IISS, 1976), pp. 22–9.

[26] See Peter Wilson's discussion of the possible impact of PGM on the amphibious tactics of the US Marine Corps (*op. cit.* in note 23).

[27] Yugoslav officials appear to have a strong interest in the new technologies and are reportedly negotiating with the United States to buy TOW ATGW. In a commentary following the military 'Victory Parade' on 9 May 1975 an observer said of the equipment displayed: 'The main impression was ... technical progress, in which, to be frank, we have not always been sufficiently strong. In this connection the first place should perhaps be accorded to missile weapons and electronic equipment designed for anti-aircraft defence, anti-tank combat and precision aiming and detection ... contrary to expectations, the development of weapons is increasingly becoming an element of defensive and people's wars, even though it had been initially oriented to the needs of aggressive wars.' (*Summary of World Broadcasts*, 13 May, 1975.)

ample manpower could use the profusion of relatively cheap and serviceable weapons (so-called 'intermediate military technology') in labour-intensive forms of defence organization.[28]

The military possibilities apparently opened up by new technologies have led observers to explore their political and strategic consequences. Perhaps the most interesting argument concerns the impact of improving NATO conventional capabilities in central Europe. It is maintained that if NATO were better able to mount a credible conventional defence, it might be able to rely less on the threat or actual use of nuclear weapons to deter or defend against attack. The attractiveness of this proposition is bolstered by the effects foreseen the promised lower collateral damage in wartime; not only could conventional forces become more effective, but attitudes towards their use could change. Europeans would thus have greater confidence that their territory could be defended without the risks attached to nuclear escalation. The nuclear 'threshold' in Europe might thereby be raised, and the West would be in a better position to consider the merits of a substantial reduction in theatre nuclear weapons, the adoption of a 'no-first-use' strategy, or the establishment of a nuclear free zone in central Europe. In the strategic nuclear sphere, the promise of strengthened conventional defences in Europe has also been seen as a means to relieve growing pressures on the 'extended' American nuclear deterrent in an era of strategic parity. For weaker states in other regions the possibility of improved defensive capabilities carries the connotation of greater deterrence, and hence greater stability. In the Middle East, South Asia or the Korean Peninsula waging aggressive wars might become more difficult, thus raising the chances of peaceful settlements to outstanding disputes being reached. Finally, it has been suggested that bolstering the confidence of states in their ability to defend themselves with conventional arms might reduce incentives to acquire nuclear weapons.

These are all intriguing claims that deserve to be treated in greater detail than is attempted here. It is important, however, to recognize that arguments concerned with the political implications of new weapons technologies must finally rest on assumptions of how these will affect the nature of future military operations. Will the new weapons have the military impact claimed for them? If so, are they likely to be exploited in such a manner that real benefits will result? These questions are examined in the next section.

III. PERFORMANCE, TACTICS AND ORGANIZATION

Operational Problems

Although the systems now entering service in several states have vastly improved single-shot kill probabilities, their efficient use depends on the successful functioning of the whole target engagement sequence, from initial identification to destruction. Within this chain of tasks there are some weak links. 'The ability of a weapon to destroy a target has often outrun the capacity to find it and range on it accurately', one writer noted in 1967.[29]

With the advent of new mobile and stand-off PGM, this gap has probably grown larger. Longer-range battlefield munitions, like guided artillery or air-launched ATGW, can destroy tanks illuminated by a laser at distances over five miles, but it is not yet clear how targets will be reliably designated during heavy fighting. Some analysts have suggested that in the future most front-line troops will act primarily as observers for stand-off weapons placed well to the rear. But the difficulty of locating and designating targets, particularly in dispersed and fluid battlefield conditions, may make it impossible to exploit the potential of PGM fully. Forward-deployed spotters and designators will have to operate in bad weather and under heavy suppression fire, and in hilly and thick terrain the lack of intervisibility between designator and target will hamper operations. Laser designation systems, moreover,

[28] See G. Pauker, S. Canby, A. R. Johnson and W. Quandt, *In Search of Self-Reliance: US Security Assistance to the Third World Under the Nixon Doctrine* (Santa Monica, Calif.: Rand Corporation, R-1092-ARPA, June 1973), pp. 60–61.
[29] Kenneth Hunt, *The Requirements of Military Technology in the 1970s*, Defence, Technology and the Western Alliance No. 5 (London: IISS, September 1967) p. 11.

will not work at night or in poor visibility and laser-equipped observers can be detected and will be vulnerable to enemy fire. The designation function will also present special problems at sea.

Command and control poses other problems. As we have seen, the trend in communications improvement is towards providing integrated, sector-wide command capabilities, able to collect information, assign target priorities and allocate strikes rapidly. This seems sensible in theory, because it enables area commanders to use resources efficiently by responding to threats according to a scheme of priorities. But in practice command-and-control centres must be able to survive an opponent's effort to destroy them or render them ineffective by jamming or interference. Systems such as AWACS, for air-defence integration, Aegis, for the co-ordination of multi-ship fire control, and fire-direction centres for artillery are lucrative targets. When command and control is centralized, an opponent can hamper effective communications by concentrating his efforts on a few important targets.

A worse problem is the threat of overburdening existing and projected command-and-control capabilities. As these have grown, so have the demands placed upon them, and the capacities of both men and equipment appear to be stretched by new requirements.[30] A new generation of longer-range anti-tank missiles, like the US army's Hellfire, is now under development, but the problem of co-ordinating multiple missile attacks against multiple targets remains to be solved. This command-and-control difficulty is perhaps most vividly demonstrated by what has been called the air-space management problem in central Europe, where in time of conflict in the mid-1980s the sky could teem with a variety of NATO helicopters and close-support, deep-strike and interceptor aircraft as well as long-range and mini-RPV. The task of co-ordinating such a

volume of air traffic, while maintaining a full range of air defences against enemy aircraft, will be daunting. This was shown during the 1973 Middle East War, when Egypt was able to deploy an effective, low-, medium- and high-altitude ground-based air-defence network, but only at the expense of greatly restricting her air force operations.[31]

More important than existing and potential target-acquisition and command-and-control deficiencies are the limitations of the most impressive of the new technologies: precision and remote guidance. Despite their performance in tests and in limited-combat situations, present-generation PGM suffer a number of operational constraints. Advanced infra-red and microwave sensors are being developed which will be able to operate at night or in poor weather, but for the rest of this decade most PGM will need clear daylight to function properly. This may not be very significant for operations in the Middle East, where daytime surface visibility and ceiling are generally good, but it is most important for central Europe, where there is darkness or bad weather over 80 per cent of the time in winter. Moreover, even in clear daylight PGM are unlikely to perform as required in urban and industrial concentrations, where extraneous energy sources and obstructions will degrade their performance.

Countermeasures present another difficulty. This area has only just begun to receive attention, but several techniques for interfering with or deceiving tracking and guidance equipment are available. This is not in itself enough to suggest that PGM will be unable to fulfill their assigned functions – potential countermeasures may turn out to be too expensive, or themselves vulnerable to simple counter-countermeasures – but there are several possible countermeasures to existing PGM guidance technologies, as Table 3 shows. One of the most vulnerable components of RPV and some PGM guidance systems is the data link that enables a weapons operator to guide missiles toward their targets. A data link's vulnerability

---

[30] See Maj.-Gen. Thomas M. Rienzi and Lt.-Col. Raymond Ketchum II, 'Command and Control in the Army', Signal, March 1975. These authors warn against over-reliance of data processing in command and control and argue that the central problem is not collecting and transmitting information, but synthesizing for the decision-maker. Noting that command and control is the major constraint on large-scale RPV operations, Under Secretary of the US Air Force James W. Plummer accused the service and industry of being 'over-optimistic about the state of technologies available' (see 'Temper RPV Enthusiasm', Aviation Week and Space Technology, 23 June 1975, p. 7).

[31] Ironically, the co-ordination problem can be exacerbated by the opponents command-and-control capabilities. In the 1973 Middle East war, large numbers of advanced systems were crowded into a relatively small area, and communications on certain frequencies were overcrowded. This led to several command-and-control failures by both sides, including the shooting down of friendly aircraft.

to jamming is largely determined by the width of its transmission band, and wide-band data links are necessary for transmitting video and infra-red information.[32] Unless counter-countermeasures are built into these systems, therefore, the ease with which they can be jammed may grow in proportion with the sophistication of stand-off guidance and control.

Table 3: Potential PGM Countermeasures

Countermeasure	Guidance Techniques
Smoke	Manual and semi-automatic wire guidance, infra-red tracking and homing, laser homing, electro-optical contrast seeker.
Fog	Laser designation and homing.
Camouflage	Electro-optical contrast seeker.
Flare	Infra-red tracking and homing.
Arc light or laser	Infra-red tracking or homing, laser designation and homing, electro-optical contrast seeker.
Radio link jamming	Command guidance, beacon position fixing.

Much simpler countermeasures, such as the extensive use of smoke and camouflage, could provide cheaper and easier ways to defeat PGM on the battlefield. Smoke alone could be used to defeat both manually-guided and semi-automatic ATGW, and lasers or high-powered arc lights could also be used to distract the latter's tracking sensors. Indeed lasers might be used to blind a variety of systems employing optical or electro-optical guidance. Flares or other heat decoys could be used to defeat infra-red homing sensors, and the generation of water vapour (fog) would reduce the effectiveness of laser-assisted weapons. The most basic form of protection, however, comes not from countermeasures but from design improvements that will enhance the ability of armoured vehicles or aircraft to survive against PGM – for instance, the reduction of aircraft heat and radar signatures by shielding engine exhausts and by using construction materials of low radar reflectivity, or modifying the construction of armour to resist shaped charges.[33]

Offensive–Defensive Interaction

Despite the important operational questions that remain to be solved, it does seem possible to draw the tentative conclusion that, taken together, the new technologies will make high-value systems, like strike aircraft, tanks and major surface warships, more vulnerable in the coming decade, particularly when operating in high density formations. Thus it is tempting to conclude that the defence stands to gain over the offence. But in determining what this means for the defence of Western Europe, the balance of power in the Middle East or the military viability of smaller states, it is essential first of all to distinguish between political and military concepts of offence and defence.

Israel, for example, can be seen as politically on the defensive (because of her position in Middle East disengagement diplomacy, the pressures applied by the United States or votes at the United Nations), but she has adopted an offensive military strategy – vividly illustrated by the deep-penetration air strikes and armoured advances that characterized the 1967 war and latter part of the 1973 war. Therefore, unless Israel fundamentally altered the offensive character of her theatre-wide strategy, it might be argued that developments reducing the survivability of armoured and air forces will work to the military advantage of her adversaries.

But this is prejudging the issue. Military concepts of the offence and defence must be further subdivided into strategic (theatre-wide) and tactical (local) components. Disregarding the issue of whether NATO's political orientation is offensive or defensive (there appear to be elements of both), the Alliance has adopted a defensive military strategy whose primary aim is to resist a Warsaw Pact attack. It is also clear that the size, deployment and doctrine of Warsaw Pact forces reflect an emphasis on 'massive, rapid offensive operations at the theatre

---

[32] A congressional study of the TV-guided *Condor* stand-off missile, now under development by the US Navy, maintains that the data link can be detected and jammed, and that to reach its target the *Condor* will need penetration aids, such as jamming, decoys and anti-radiation missiles. 'Letter Report to Secretary of Defense on Production Decision on the Condor Weapon System' (General Accounting Office, November 1975).

[33] The use of composite (metal and polypropylene) armour to counter or withstand the effects of small shaped charges looks particularly promising. See Richard Ogorkiewicz, 'The Future of the Battle Tank', in Kemp, Pfaltzgraff and Ra'anan (eds), *op. cit.* in note 23; see also reports of developments in British tank armour (e.g. Henry Stanhope, 'Britain invents tough new tank armour', *The Times*, 18 June 1976).

level in Europe'.[34] Thus, on the theatre level, by enhancing NATO's capabilities to defend against a Warsaw Pact armoured attack, it would appear that the new technologies will favour the West.

When the tactics, deployment and organization of Western forces are taken into account, this judgment must be greatly qualified. Despite a defensive doctrine on the theatre level, existing NATO doctrine calls for a 'dynamic' mobile defence, which would include a variety of offensive operations, including local brigade- and division-level armoured counter-attacks and deep-penetration air raids over enemy territory. In the maritime sector, amphibious operations and strike missions by carrier-borne aircraft are planned. As several observers have pointed out, Western forces are organized and equipped for these offensive missions.[35] Similarly, not all Warsaw Pact forces are configured for continuous *blitzkrieg* combat; those of the East European states in particular, are not. Even Soviet Army doctrine, which places such an extreme emphasis on offensive operations, admits the need for defensive tactics on the local level. In this situation, it becomes far more difficult to measure the impact of the new weapons beyond concluding that, in Europe as elsewhere, while ground may become more difficult to seize, once gained, it will be almost as difficult to recover.

In this context it should be noted how the proponents of the new weapons equate the offensive with a concentrated armoured attack. Such an offensive may be blunted by dispersed, PGM-equipped anti-tank units, but anti-tank

strongpoints will have problems to face in defending against combined infantry and artillery operations. Israeli tactics in the latter days of 1973 Middle East war demonstrated that artillery fire and infantry sweeps preceding the use of armour are an effective means of countering dispersed anti-tank units, and this concept is well understood by Soviet tacticians. The optimal defence deployment against what J. F. C. Fuller called the 'shock' offensive (fast-paced armour operations) is unsuited for defending against the 'missile' offensive of slower-moving artillery fire and infantry sweeps: defensive tactics must be tailored to the evolving offensive threat.

In the PGM era, then, an attacker might exploit the defensive advantages of the new weapons on the tactical level in pursuance of an offensive strategy on the theatre level. Because concentrated armoured thrusts may become less effective in the light of new technologies and defensive dispositions, the attacker might abandon these tactics in favour of surprise raids designed to deny his opponent sufficient time to establish defences in depth. PGM-equipped forces would seize territory and form their own defensive strong points, while longer-range PGM could be used to strike at command centres, supply stores and air bases. In this way the attacker could use the defensive capabilities of the new technologies for offensive ends.

It could be questioned whether specific opponents, such as the Warsaw Pact armies, have the equipment or the doctrine to undertake this type of operation, but it is important to recognize that the new vulnerability of tanks and aircraft does not inevitably work against the aggressor; pre-emptive missile strikes, coupled with rapid advance by PGM-equipped units, might be an effective offensive instrument in central Europe, particularly against forces still wedded to tanks as a primary means of defence. On the flanks these tactics might entail fewer military risks and greater political dividends. In other regions, the potential for local surprise attacks appears even greater, because the capacity for successful defence of newly-seized territory could influence military intentions in the Sino-Soviet border dispute, in the Indo-Pakistani conflict over Kashmir, in territorial claims in the Gulf and, of course, in the Middle East.

In summary then, many of the arguments of the proponents of the new technologies appear

[34] Jeffrey Record, *Sizing Up the Soviet Army* (Washington, DC: The Brookings Institution, 1975), p. 33. While there are differences among commentators over the conditions that might provoke the Soviet Union to initiate military operations in Europe, there is a general consensus that in the view of Soviet doctrine 'the land battle would involve swift armoured thrusts by tanks and mechanized infantry aimed at penetrating deep into enemy territory'. See Trevor Cliffe, *Military Technology and the European Balance*, Adelphi Paper No. 89 (London: IISS, 1972), p. 34.
[35] For a discussion of the emphasis placed on deep-penetration air operations in the design of aircraft and the assignment of missions by the US Air Force, see William White, *op. cit.* in note 1, pp. 73–76. The emphasis on offensive 'expeditionary-style' operations can also be seen in the doctrine and equipment of American land forces in Europe. See Canby, *op. cit.* in note 23, pp. 15–22, and Richard D. Lawrence and Jeffrey Record, *US Force Structure in NATO: An Alternative* (Washington DC: The Brookings Institution, 1973), pp. 27–44.

too simple. The state of technology does condition the nature of the offensive–defensive duel, but it will not necessarily determine the outcome. An offensive that is not adapted to new military capabilities may well be checked by defensive forces that are, but the reverse is also true. To talk of 'defensive' weapons (as opposed to weapons that lend themselves to defensive operations on the tactical level in pursuit of either defensive or offensive ends) is generally unhelpful.[36] This means that the likely military impact of the new technologies can only be measured in a specific context. The case of NATO and the Warsaw Pact is more closely examined in the next section.

Technology and the European Balance
Acquisition by the Adversary

In examining the likelihood of the new weapons altering existing offence–defence relationships, the foregoing paragraphs assumed they were adopted roughly symmetrically. In the NATO–Warsaw Pact case, this must be questioned. When advocating the widespread deployment of the new systems by NATO forces, several observers have pointed to Western technological prowess and apparent Soviet weaknesses in basic and applied research in order to argue that Warsaw Pact forces will not have equipment comparable to that entering NATO inventories for at least another decade. In most areas of Research and Development this seems to be true.[37]

In brief, the Soviet Union is not known to have introduced either electro-optical or laser-guided weapons, and the United States is said to enjoy a considerable edge in air-to-surface

guided weapons.[38] However, she is deploying a new family of infra-red-guided air-to-air missiles, and an advanced radar-guided version of the AA-6 *Acrid* is reported to have entered service with Pact interceptors. A new radio-command air-to-surface missile, the AS-6 *Kerry*, said to have a range of 800 km, is under development for deployment aboard the *Backfire* bomber,[39] and a radio-beacon positioning system, similar in operation to LORAN, is also being deployed in Eastern Europe.[40] It is interesting to note that the deployment of the first Soviet fighter apparently designed specifically for ground-attack, the Su-19A, suggests that air-to-surface PGM will be given greater priority in Soviet research and development during the next decade.

The evidence of the 1973 Middle East conflict shows that the Soviet Union's technological inferiority in anti-tank missiles is far less marked than in air-delivered munitions, and that in one area – air defence – she has deployed a wider variety of more capable systems than the West. Since the October war, two new battlefield SAM systems, SA-8 and the SA-9, have been introduced to Soviet forces, and the former is said to have an electro-optical tracking capability. In addition to three (rather old) ATGW types, Soviet infantry is equipped with a variety of target-acquisition aids, including image intensifiers and infra-red scopes.[41]

Soviet advances in anti-ship missile capabilities are well known. Although present Soviet anti-ship missiles are less accurate than systems like *Exocet* now in service in the West (particularly in the case of over-the-horizon types), a new generation of naval systems like the SS-N-12

---

[36] Colin Gray has noted that even the most 'non-offensive' of weapons have historically been employed for offensive ends. In feudal warfare castles were occupied and constructed in order to hold territory just gained, and in modern warfare atomic demolition mines could be exploited by attacking forces to deny counter-attack corridors to an adversary on the defensive. See 'New Weapons and the Resort to Force', *International Journal* (Toronto), Spring 1975. For this reason, attempts to define offensive and defensive weapons in the context of arms-control negotiations have always encountered difficulties. On this point, see Geoffrey Kemp, 'Arms Control and Third World Conflicts', *International Conciliation*, March 1970.
[37] According to the US Department of Defense, the United States has advantages in several basic areas of technology: inertial and sensor-aided guidance, integrated circuits, computers, turbofan and ramjet engines, aerodynamics, and composite materials. In a few technologies – high-pressure physics, welding and high-frequency radio – the

Soviet Union is reported to have the lead. See *Statement by Malcolm R. Currie, Director of Defense Research and Engineering, to the 94th Congress*, February 1976.
[38] Gen. George Brown, *op. cit.* in note 3, p. 106. However, Secretary of Defense, Donald Rumsfield, has noted 'the Soviets are developing a variety of new air-launched weapons including a family of tactical air-to-surface missiles and bombs', (*Annual Defense Department Report, FY 1977*, p. 127) and it is reported that an anti-radiation air-to-surface missile is under development in the Soviet Union. See *Nouvelles Atlantiques*, 12 June 1976, p. 3.
[39] Europe's New Generation of Combat Aircraft, Part I: The Increasing Threat', *International Defense Review*, April 1975.
[40] *Aviation Week and Space Technology*, 22 November 1971, p. 38.
[41] See Capt. Eugene D. Betit, 'Soviet Preparation for Night Combat', *Military Review*, March 1975.

cruise and the SS-N-13 ballistic missiles do have improved guidance systems and longer-ranges than their Western counterparts. In space, while there is no evidence that the Soviet Union is developing a satellite position-fixing system, several advanced surveillance and communication systems were launched in 1975, including synchronous-orbit and maritime-reconnaissance satellites.

In general, then, it is difficult to gauge the extent and permanence of the Soviet technological gap. But one should recognize that, if and when the Soviet Union deploys new weapons types, she could be able to deploy them more efficiently, by virtue of greater standardization within the Warsaw Pact, and in greater numbers, because of generally larger production runs. This is especially true of PGM, because she has more potential delivery vehicles (tactical aircraft, surface-to-surface missiles, artillery tubes and rocket launchers).

What would be the impact of the Warsaw Pact deploying the new systems in the same or greater numbers than NATO? The crude model, stressing NATO's defensive mission and the defensive orientation of the new weaponry, suggests that if NATO acquired sufficient anti-tank and air-defence weapons to counter Soviet aircraft and tanks, the Pact's possession of a similar capability would have little impact on the overall balance. This analysis is appealing because it borrows from the distinction between offence and defence used in the sphere of strategic weaponry. But, as discussed above, the interaction of forces in land warfare is far more complex; in combined arms operations where suppression and manoeuvre are elements both of the offence and defence, it is unclear, in the abstract, which side is more likely to benefit from PGM and associated technologies. Because battlefield and (more obviously) longer-range PGM can be employed for offensive purposes, the effect of their deployment by the Warsaw Pact could be far greater than is generally assumed, especially at a time when certain NATO missions (like deep-penetration air strikes) appear less viable rather than more so, in the light of new weapon developments.

Major factors in examining the impact of Soviet acquisition are the general character of NATO and Warsaw Pact forces and the nature of the targeting patterns each side presents to the other. Here, the frequent observation that NATO relies on fewer but more capable systems (high cost, low attrition forces), while the Pact has more but less sophisticated and more expendable equipment (low cost, high attrition forces) suggests that NATO could be at a more severe disadvantage in a 'one shot, one kill' era than at present. This is given added force by NATO's present dependence on a large and vulnerable logistics base. The same pattern is also apparent at sea, where Western naval power is concentrated on a declining number of high-value platforms, while Soviet construction shows an emphasis on greater numbers of less capable vessels. Thus, if the widespread deployment of PGM leads to greater decentralization of weapons release (and hence higher consumption rates) and greater kill probabilities (and hence higher attrition rates), the advantage may go to the side with larger forces-in-being.

Adaptation by the Adversary
It was suggested above that, given existing NATO and Warsaw Pact force structures and dispositions, symmetrical deployment of the new weapons could in some instances work to the advantage of the East. There may also be cases where the Pact, in attempting to adjust to NATO's deployment of new systems, could not only nullify their impact but pose new and unfamiliar threats.

Is large-scale Soviet adjustment to NATO acquisition of a new generation of weapons likely? Despite the operational flexibility attributed to Soviet forces, it is clear that the land forces are presently armed, organized and trained to fight the 'short war': a continuous and rapid armoured advance aimed at overrunning the whole of the central region. The ability of new weapons to inflict rapid and high attrition rates upon exposed targets in intense exchanges of fire therefore seems to make them well-suited to counter this threat. However, the rapid rates of advance postulated in Soviet doctrine assume the continued willingness of commanders to commit large numbers of combat vehicles into battle; some writers have speculated that, in the longer term, PGM will slow the pace of offensive operations by inhibiting the early and liberal use of tanks, mechanized infantry and aircraft. The need to inhibit PGM fire, meanwhile, will call for large quantities of suppression fire. The result is

60

that 'combat will shift towards a struggle of attrition in which peripatetic infantry will play the central role in many tactical offensive operations'.[42] This seems an important argument, for it assumes that the would-be attacker would 'learn', in the face of growing defensive firepower, accepting slower movement and the adoption of a force mix that placed less emphasis on armour and gave a greater role to infantry and artillery fire in an offensive.

The crucial issue is whether the Warsaw Pact would or could undertake the changes necessary to counter NATO deployment of a new generation of anti-tank weapons. A common assumption is that Soviet doctrine is wedded to the idea of rapid advance and that, as a result, the Warsaw Pact armies are unlikely to de-emphasize the role of tanks in the land-force mix.[43] But, while it is true that Soviet doctrine continues to stress the importance of rapid and continual advances, the army has demonstrated a capacity to adapt to new conditions. In moving from an emphasis on closely-packed armour formations to more widely dispersed forces, Soviet doctrine did respond rapidly to Western deployment of theatre nuclear weapons in the early 1960s, and in recent years there seems to be a shift of emphasis from tanks towards bolstering the capabilities of artillery and mechanized infantry. Thus, a new generation of self-propelled artillery has been deployed, as well as thousands of armoured personnel carriers for motorized rifle divisions (BMD and BMP). These changes reflect a number of considerations – including, as Phillip Karber has argued, a deep appreciation of the threat posed by ATGW.[44] However, Western PGM deployment is unlikely to cause the abandonment of the general notion of rapid, continuous advances; means of maintaining this capability in the face

of Western ATGW are likely to be stressed instead. Several solutions are possible (and appear to be under discussion by Soviet commanders), including the use of nuclear weapons and/or conventional artillery to suppress anti-tank defences and the use of surprise raids (of the type discussed on p.58) to allow Pact forces to exploit the benefits of ATGW for their own ends.[45]

It is not my object here to predict how the Warsaw Pact will adjust to Western deployment of ATGW and other PGM – only to point out that the different choices open to Soviet planners have widely different implications for the nature of a potential conflict in Europe. In the short run Soviet options appear to be narrowed by a doctrinal preference for the 'short war' and by the sheer difficulty of radically changing the mission and organization of a bureaucracy the size of the Warsaw Pact. In the longer term the Warsaw Pact has the front-line equipment and a logistics and support base that could be used in a strategy that envisaged a more prolonged conflict.[46] If the Warsaw Pact eventually turned to a strategy that emphasized PGM suppression and accepted a slower rate of movement, the impact of such a change could be profound. Consider, for example, the incidence of damage to civilian structures and population on and off the battlefield; precision guidance and improved target acquisition promise to lower collateral damage, but if the Warsaw Pact relied on extensive infantry sweeps and artillery fire to suppress defences in depth, higher levels of collateral damage could result. This would be particularly true if militia-type forces used towns and villages as defensive strong points, as some observers have suggested.[47]

---

[42] Wilson, op. cit. in note 23, p. 23. Martin Van Creveld argues that future battles will 'degenerate into slugging matches going on interminably'. See Military Lessons of the Yom Kippur War: Historical Perspectives (Beverly Hills: Sage Publications, 1975), p. 39.

[43] Col. Edward Atkeson, for instance, argues that continued Soviet emphasis on tank operations reflects 'battleship thinking', (see 'Is the Soviet Army Obsolete?', Army, May 1974). Jeffrey Record similarly claims that the Western debate over the implications of ATGW for the vulnerability of armoured forces has had no parallel in the Soviet Union (Lawrence and Record op. cit. in note 37, p. 48).

[44] Phillip Karber, 'The Soviet Anti-Tank Debate', Survival, May/June 1976, pp. 105-111.

[45] These options are discussed in detail by Karber, who notes that – because Soviet concern over ATGW appears to be directed towards the survivability of BMP more than tanks – one option under discussion is to put even more emphasis on tanks.

[46] See Graham Turbiville, 'Society Logistic Support for Ground Operations', RUSI Journal, September 1975, pp. 63-9.

[47] The West European central region is the most urbanized area in the world. The growth of the so-called Rhine–Ruhr–Dutch Randstand complex since World War II has ambiguous implications for Western defence: on the one hand, it has reduced the open area available for armoured operations, making it progressively more difficult for Warsaw Pact armies to make rapid advances by circumventing built-up areas. At the same time, it increases the likelihood of severe collateral damage in conventional

The new technologies do appear to be well-suited to present-day NATO requirements. But Western acquisition of, and Eastern adaptation to, the new weapons could have the paradoxical effect of making NATO forces better equipped to defend against an armoured attack on the North German plain, but increasing the Alliance's exposure to other threats.

Costs and Mission Priorities

If the issues raised in the preceding three sections are resolved in favour of exploiting the new technologies, the even more complex problem arises of how these systems might be optimally deployed, given prevailing force structures and organization. As we have seen, this is primarily a question of the compatibility of the new technologies with existing roles and missions. Ultimately, however, it is related to costs. The funds that states will be able to spend on the new systems will be constrained by efforts to maintain existing capabilities, regardless of whether these are still important in the light of new developments.

Those who maintain that the new systems will enhance the defensive posture of NATO and smaller states make the important assumption that their relatively low cost will enable large numbers to be procured. But there are different approaches to measuring the financial impact of the new weapons, each of which raises special questions.

Procurement costs

In discussing the economic attractiveness of the new systems, acquisition costs of less than $10,000 are frequently cited for battlefield anti-tank missiles like the *TOW* or shoulder-launched SAM. These figures refer to cost per round and neglect the launcher (which costs over $20,000 in the case of *TOW*). Far more important, however, is the much higher cost of more sophisticated and longer-range PGM and RPV; the

Harpoon cruise missile costs $600,000 while *Firebee*-type combat RPV will cost over $1 million each. Another factor is the cost of the weapons platform from which the systems are launched. Many battlefield weapons are man-portable but, in an effort to provide greater mobility and protection, SAM and anti-tank missiles are being deployed aboard combat vehicles, and the cost of the British tracked *Rapier* system, including SAM launcher and carrier, for instance, is $1·25 million. In calculating the cost of air-launched munitions, the price of the stand-off platforms – helicopters or fixed-wing aircraft – must also be considered.

A more neglected area in spending discussions is the cost of the target-acquisition and command-and-control equipment that in many cases is vital to the performance of new weapons. Often less visible than other areas of procurement, projected command-and-control improvements will entail substantial expenses. The proposed fleet of 32 AWACS aircraft to manage the air battle over Europe would cost about $2 billion, and the procurement cost of the 24-satellite Global Positioning System also runs into billions.

Replacement Savings

The costs of these systems cannot be considered in isolation but must be judged in terms of the potential savings accruing from substitution. For instance, it has been argued that a 600-mile-range tactical cruise missile equipped with terminal guidance could provide a more economical alternative to combat aircraft, especially over heavily defended areas, when subsidiary mission costs (fighter escorts, rescue and electronic warfare aircraft) are taken into account.[48] Even for permissive air environments, it has been estimated that the relative mission costs of high-performance aircraft vary from twice to ten times that of RPV.[49]

These cost comparisons are impressive, but one should bear in mind that much of the new technology is seen not in terms of replacement options but in terms of enhancing the role of existing systems. Cruise missiles and RPV may ultimately replace combat aircraft, but it is more

and/or nuclear conflict, and this has led the West German government to re-emphasize the importance of a forward defence strategy. See the White Paper, *The Security of the Federal Republic of Germany*, (Bonn: January 1976, p. 25) Paul Bracken has speculated that the difficulties of by-passing urban areas and the improvement of NATO defences in open areas could ironically serve to make cities more attractive targets for attack by the Warsaw Pact. See 'West European Sprawl as an Active Defense Variable', in R. Huber, (ed.), *Military Strategy and Tactics* (New York: Plenum Press, 1975), pp. 219–230.

[48] Richard L. Garwin, 'The Shape of Future US Military Forces' (unpublished paper), 4 July 1972, p. 6.
[49] This assumes an attribution level of 0·01 per sortie. See Kent Kresa and Col. William F. Kirlin, 'The Mini-RPV: Big Potential . . . Small Cost', *Astronautics and Aeronautics*, September 1974, pp. 48–62.

likely that over the next decade they will be used to maintain the effectiveness of manned systems. This process can also be seen in the area of battlefield systems, where substitution will be limited by pressures to obtain more effective force mixes. (During the 1973 Middle East conflict Arab forces used mobile SAM systems not to supplant anti-aircraft cannon fire but actually to complement Soviet-supplied air-defence guns.) A similar situation exists, with anti-tank weapons, guns being more effective at shorter ranges and missiles more accurate at ranges over 1000 metres.

Life-cycle costs

The research, development and procurement costs of new equipment are often far less than those of operating and servicing the systems over their useful lives. This is particularly true of more sophisticated systems that require large numbers of skilled maintenance personnel. The growing need to procure systems that are economical in their use of skilled manpower – felt especially keenly in Western states, where personnel costs are rising rapidly – can be met in several ways: modular design enables non-specialists to replace defective components; the use of simulators can reduce the cost of training with expensive munitions; and, perhaps most important, systems can be designed to operate with the minimum number of men (the *Lance* missile launcher, for instance, needs only half as many crewmen as the systems it is replacing).

Organizational Adjustment

However, the general trend towards capital-intensive equipment raises important questions of its own. Some states, like the United States, Germany and Britain, are well placed to exploit such an approach, but Turkey, for instance, lacks the necessary technological infrastructure and at the same time is not subject to the manpower-cost squeeze bedevilling her more advanced Alliance partners. For lesser-developed states, like Iran, the procurement of capital-intensive systems is even less necessary. As a result, a different approach to weapons design is needed: one that capitalizes on a plentiful supply of manpower and on intermediate technologies. The resulting types of 'light-weight, relatively simple but highly effective weapons', like man-portable air-defence and anti-tank missiles (which have also been discussed in relation to bolstering the

potency of European reservist and militia forces[50]) do, however, pose operational problems: the difficulty of identifying friend and foe, and the threat of their theft and use by terrorists and criminals. The approaches to weapon design, whether capital- or labour-intensive, need to be tailored to specific national contexts.

With these considerations in mind, one must ask whether armed forces will be willing to make the economic and organizational adjustments necessary to exploit the new technologies to the full. It was noted earlier that, despite a defensive strategy on the theatre level, NATO ground, air and naval forces continue to place emphasis on offensive tactics and organization. By and large, as Steven Canby, Robert Komer and others have argued, NATO forces are not configured for the dispersed and decentralized style of combat that seems best suited to the new weaponry and the existing character of the Warsaw Pact threat. This is a problem that Alliance members may share with other states who are beneficiaries of Western military assistance and have organized their forces on Western models. American army divisions, with their large support complements and comparatively low densities of anti-tank weapons and artillery tubes per unit, are a good example of this.[51] Perhaps a better one is offered by American tactical air forces, which plan to rely on advanced aircraft to strike at interdiction targets and air bases deep in enemy territory. Whether this capability should have the high priority it now enjoys is a separate and

---

[50] See Kenneth Hunt's discussion of militia forces and technology in *The Alliance and Europe: Part III, Defence with Fewer Men*, Adelphi Paper No. 98 (London: IISS, 1973), pp. 31–3, and Horst Mendershausen, *op. cit.* in note 23, pp. 13–14. There is an important distinction between the ability to design, develop and procure a system and the ability to maintain and operate it. An advanced ATGW incorporates more sophisticated technology than a tank of Korean war vintage but, while the latter requires experienced tank crews and mechanics, the former demands little training and virtually no maintenance. This is important for states that do not possess the capability to produce advanced systems but have the money and manpower to operate them in large numbers.

[51] While the US Army plans to purchase 30,000 *TOW* ATGW through 1977, it is presently using them to replace the 106mm recoilless rifle in infantry units on a one-for-one basis. This will keep the density of major anti-tank weapons per unit unchanged and leave half again as many soldiers per anti-tank weapon as in Soviet units. Also, a far higher proportion of Soviet anti-tank weapons are protected by armour.

complicated issue revolving around questions about the likely duration of war in Europe and whether NATO could maintain air superiority over the battlefield without extensive counter-air operations. The important point here is that much of the money and effort now directed towards exploiting new weapons technologies has been devoted to maintaining the effectiveness of deep-strike aircraft. However, new developments suggest that the enhanced performance of systems on and over the battlefield (SAM and air-combat fighters like the F-16) may make the deep-penetration capability less important and that, insofar as the capability is still needed, other and cheaper means of providing it (cruise missiles and RPV) will be available in the 1980s. The same may also be true for the close air-support mission. A new generation of terminally-guided stand-off weapons should enable air forces to deploy less expensive helicopter and fixed-wing platforms. It is even possible that land-based systems, such as precision-guided-artillery, could obviate the need for close air-support aircraft in most situations.

Similar arguments can be applied to the US Navy's aircraft carriers, whose size and sophistication is largely determined by the requirements of the mission of projecting sea-based air power ashore. If this mission were down-graded in response to the increasing problems of penetration, or if attack aircraft were replaced by a new generation of missiles, the Navy could move to smaller, cheaper (and consequently more numerous) 'midi-carriers', operating aircraft which were suitable for less demanding missions. At present, however, the Navy's technological enthusiasm continues to be directed towards ensuring the survivability of the large, multi-purpose carrier as a means of maintaining the viability of the projection mission.

A re-appraisal of mission priorities and equipment is needed in order to avoid funds for exploitation of the new technologies being frittered away by efforts to perform less relevant missions more effectively (or simply to maintain effectiveness when an adversary acquires advanced equipment). If aircraft are increasingly vulnerable to air defences, the right answer does not seem to be to fit more sophisticated defence-suppression aids to fighter-bombers costing over $15 million and to procure electronic-warfare support aircraft, like the EF-111A. It would make more sense to rely instead on a mix of weapons over and on the battlefield to establish local air superiority and on longer-range missiles for deep-penetration strikes.

Because some new weapons may challenge established roles and missions, it is not only possible that certain technologies, will be inappropriately exploited but there is the danger that they won't be exploited at all. As Graham Allison and Fred Morrison have noted, the US Air Force resisted the deployment of air-launched PGM for several years, because they were seen to imply a reduction in aircraft force levels and in the service budget.[52] New technological developments may not only be neglected because they challenge existing missions and equipment, but because they threaten other new options. Terminally-guided rocket and artillery systems may prove to be far more cost effective than cruise missiles or RPV in shorter-range, shallow interdiction roles, but the services may not investigate their feasibility for fear of jeopardizing more elaborate programmes.

In the case of NATO, another issue closely related to costs and mission structure is equipment standardization and interoperability. In Part I it was noted that NATO members are currently engaged in developing eighteen different anti-tank weapons. These systems reflect different doctrinal preferences, and in some cases are designed for use in different conditions and on different platforms. All the same, it is clear that such proliferation reflects unnecessary duplication of effort, which (through higher unit costs) will ultimately be reflected in the number of new systems the Alliance will be able to afford to deploy in the coming years.[53] It has been

[52] 'The Policy Process and Institutional Constraints', in Holst and Nerlich (eds.), op. cit. in note 18.
[53] NATO (including France) has 105 different tactical missile systems in service or under development, compared with 25 for the Warsaw Pact:

	Naval (sea-to-sea or sea-to-air)	Army (surface-to-air)	Anti-tank	Air-to-surface	Air-to-air
NATO	26	17	24	20	18
Warsaw Pact	10	5	3	2	5

SOURCE: European and Atlantic Co-operation in the Field of Armaments, Assembly of Western European Union, Document 689 (Lemmrich Report), 1 December 1975, Appendix I.

stressed, moreover, that effective exploitation of new guidance technologies will depend on the more efficient target acquisition and command and control. In a more decentralized battlefield environment, more compatible targeting and communications equipment become essential, and proliferation of different equipment fulfilling the same role becomes a serious problem.

The efficient introduction of a new generation of weapons into the arsenals of individual states, and into NATO generally, presupposes changes in patterns of defence organization and procurement which some observers have supported on their own merits. However, when arguing for the restructuring of forces into more but smaller units, providing higher ratios of 'teeth to tail', these observers have criticized proponents of the new technology for ignoring deficiencies in force organization. But restructuring and the deployment of new weapons should not be viewed in 'either, or' terms. Acquiring these systems will strengthen the case for the organizational changes advocated, and the potential offered by new technologies is crucial to any restructuring designed to give NATO a more credible defensive capability.

The same is true in the area of standardization. This issue has received renewed attention, and proposals have been made to establish European and Alliance-wide development and procurement programmes so that economies of scale could yield a more efficient return on investment. These ideas seem to assume technological and doctrinal agreement among NATO members but the promise of the new weapons (and the need to acquire them in large numbers) does make multinational procurement schemes attractive; also, increased standardization will make it easier to achieve the overall increase in capabilities that the new technologies offer.

IV. POLITICAL AND STRATEGIC PERSPECTIVES

For the time being, a conjunction of technological, economic and organizational factors appear to offer states, alone or in alliance, the ability to better provide for their defence, a situation that led some to complain that the only commodity that is presently lacking is political will. This assessment, however, ignores further political and strategic factors that are addressed below.

NATO Politics and Nuclear Strategy

It is probably inevitable that some Europeans will survey the wide range of weapons now under development and the arguments supporting their deployment and suspect that the United States is only attempting to substitute technology for manpower in Europe – asking Europeans to function as 'foot soldiers' while Americans (and probably fewer of them) operate a new generation of systems. This suspicion is undoubtedly reinforced by arguments that couple the deployment of new weaponry or the restructuring of forces with possible manpower savings.

It is true that the most advanced systems are being developed in the United States and are being most rapidly deployed by American forces, a fact which reflects the present imbalance in military research and development (R&D) expenditure between the United States and her major European allies (see Table 4). But while American military R&D expenditure dwarfs her partners', in comparison with earlier periods the gap is narrowing. Between 1955 and 1965 the combined military R&D spending of Britain, France and Germany was 10 per cent of the American total,

Table 4: Defence Expenditure and Military R & D
(Annual Average 1970-4 in $ million)

Country	Defence expenditure	Military R&D	R&D as % of defence expenditure
USA	78,000	8,736	11·1
Britain	7,600	840	11·1
France	7,500	1,092	14·6
Germany	9,300	451	4·8

SOURCES: Defence expenditure figures are derived from the relevant editions of *The Military Balance* (London: IISS). R&D figures are derived as follows: USA, *Infodoc Survey* IS/RTE 150774, IS/AME 200774 (Stuttgart: Infodoc); Britain, *Statement on the Defence Estimates 1970, 1971, 1972, 1973*, (London: HMSO, 1970-74, Cmd nos 4290, 4592, 4891, 4231); France, *Budget Voté de 1970, 1971, 1972, 1973, 1974* (Paris: Imprimerie Nationale, 1970-74); Germany, *Infodoc Survey* IS/AME 5 October 1974 (Stuttgart: Infodoc).

but by 1970–74 this figure had reached 27 per cent.[54] Moreover, the R&D gap should not obscure the fact that in several areas – anti-tank missiles, low-altitude SAM, anti-shipping missiles and target-acquisition aids – advanced systems are undergoing development and deployment by Europeans.[55]

Yet unless Europe increases collaboration on producing and procuring these systems, and the United States can be persuaded to buy European, it is hard to avoid concluding that in economic terms the United States stands to gain most from procurement of the new weapons, and that militarily exploitation of new technology will widen existing gaps in Alliance capabilities. Two consequences of the latter prospect might be foreseen. Firstly, the rapid introduction of advanced American equipment into the European theatre might be seen as an effort at subtle withdrawal from a direct commitment to Europe through a process akin to 'Vietnamization' – trading equipment for presence. Secondly, American monopolization of the new technologies might be viewed as an attempt to create increased European dependence upon the United States, leaving individual countries hostage to arbitrary decisions (such as the 1975 congressional embargo on military assistance to Turkey) and giving the United States an important lever over European behaviour.

Fears of this sort exist – most notably in France[56] – but they should not be exaggerated. In 1975 the United States showed a new sensitivity to these concerns, and in undertaking to purchase or license-build European equipment (the *Roland* II SAM, air-defence guns, naval patrol vessels and, possibly, a new tank) showed growing interest in a 'two-way street' for NATO procurement. Moreover, her decision to replace headquarters and support personnel withdrawn

from West Germany with two new combat brigades indicates that a 'Vietnamization' process is not at work. But, most important, continued progress in intra-European and intra-Alliance collaboration and co-operation in weapons design and production will rule out the possibility of American domination of the new weapons market.

A more fundamental problem is the American tendency to view deployment of new weapons as part of what former Secretary of Defense James Schlesinger called 'the gradual evolution towards increasing stress on the conventional components, a diminution of the threat of recourse to nuclear weapons'.[57] The possibility of nullifying offensive armoured threats by means of comparatively cheap conventional weapons is indeed attractive, especially when their use promises comparatively little collateral damage. The combined possibilities of greater firepower and lower collateral damage have led some to argue that a conventional 'defence in depth' in the central region may be not only feasible but also acceptable to those living there. However, as the questions raised in Part III suggest, the long-term implications of these weapons for conventional combat are by no means clear. A number of factors – the rate of munitions fire and consumption, the pace of movement, targeting strategies on and off the battlefield and the possible impact of countermeasures – must become clearer before claims about defence effectiveness and collateral damage can be made with certainty. One thing is clear, though: if the new weapons are seen as involving a significant chance of intense and protracted conventional war, many Europeans will greet a strategy based on their wholesale introduction with scepticism and concern.

The question of enhancing NATO's conventional capabilities, then, forms a central part of the enduring Atlantic debate on what constitutes deterrence and how risks are apportioned if it fails – in other words, how escalation will proceed and who will control it. This becomes clearer when claims about the new weapons are linked with raising of the nuclear 'threshold'. Emphasizing NATO's conventional defences while trying to rely less on nuclear weapons for deterrence and

---

[54] For a breakdown of European and American R&D figures during the 1955–65 period, see R. L. Pfaltzgraff, 'Research and Development Expenditures: a Comparative Analysis', presented at the Conference on Atlantic Technological Imbalance and Collaboration, Deauville, May 1967.

[55] For a survey of European developments in advanced guidance, tracking and target-acquisition equipment, see 'Advanced Electro-Optics Demand Grows', *Aviation Week and Space Technology*, 16 June 1975, pp. 48–54.

[56] See Marc Perrin de Brichambaut, 'Les "précision guided munitions" et la défense de l'europe', *Contrepoint*, no. 20, February 1976.

[57] James R. Schlesinger, 'Nuclear Weapons: An Outline of US Policy', (Press briefing on 1 July 1975), United States Information Service.

defence will be seen in Europe as a continuation of McNamara's effort to reduce the risk of nuclear war to the United States by delaying recourse to nuclear weapons in the event of conventional war in Europe – an effort that is understood to possess a new urgency in an era of strategic parity.[58]

The problem of apportioning risks in European defence, and its close connection with efforts to bolster conventional capabilities, can perhaps most clearly be seen in differing attitudes towards the role of theatre nuclear weapons (TNW). The suggestion that, with improved conventional defences, the Alliance could make significant reductions in the TNW stockpile must be examined against the familiar observation that on the two sides of the Atlantic they are seen to fulfil different functions. An effort to couple acquisition of new conventional weapons with reduced reliance on nuclear forces would seem to offer the advantage of dampening the threat of escalation to nuclear use if large-scale conventional war broke out. However, many Europeans still maintain that it is this threat of escalation that deters the outbreak of war. Raising the nuclear threshold might lower the threshold of conflict itself, if it were believed that extensive conventional operations could be carried out with little risk of escalation to the nuclear level.

This debate could be merely academic, however, for a central problem of NATO strategy is its assumption that the initial phase of a European conflict would not involve nuclear weapons, and that the Alliance could decide if, when and how to introduce them. A Warsaw Pact decision to use nuclear weapons at the outset, or early in a conflict, would therefore seem to make the Alliance 'risk sharing' debate irrelevant. The willingness of the Soviet Union to introduce nuclear weapons early in a conventional conflict is a contentious issue in the West, but it

should be recognized that Soviet intentions will to some extent be conditioned by perceptions of NATO nuclear doctrine and deployment. If deterrence has broken down a NATO conventional-emphasis strategy might invite the early theatre use of nuclear weapons by the Soviet Union because it appeared to reduce both the chances of Warsaw Pact conventional success and the likelihood of Western recourse to nuclear use. Correspondingly, a Warsaw Pact conventional strike would appear most likely in a situation where NATO was least able to respond in kind, because of a nuclear-emphasis 'tripwire' strategy.

This is not a new problem, of course, and Alain Enthoven and others have long argued that a conventional-emphasis strategy does not have to raise the 'spectre of denuclearization' if NATO maintains a secure residual force of theatre nuclear weapons which is large enough to deter Warsaw Pact nuclear use.[59] But the credibility of the Alliance's theatre nuclear deterrent posture is a function not only of NATO's conventional capabilities but also of the evolving conventional and nuclear capabilities of the Warsaw Pact. A good case can be made for 'modernizing' NATO's nuclear posture, through it is not immediately evident that a large reduction in the existing theatre nuclear stockpile is warranted. But, even if it is, there is a basic political dilemma: for some the prospect of gaining more control over the escalation process by replacing nuclear weapons with improved conventional munitions is the most attractive aspect of the new weapons technology; but relying on this argument to justify deployment of the new systems could cast doubt on the American nuclear commitment – thus stimulating European efforts to acquire independent escalation options.

According to one school of thought, the solution is to abandon the preoccupation with thresholds and exploit the new technologies for both nuclear and conventional munitions, in order to increase both the range of available options and uncertainty over the response any Warsaw Pact initiative would meet. Dampening incentives for nuclear escalation seems to be the most important virtue of the new technologies from one point of view, but their capabilities suggest other alternatives, including equipping

---

[58] De Brichambaut, for instance, argues that PGM constitute the European aspect of a return to the 1960s 'flexible response' debate. 'For the more distrustful Europeans, PGM are stained with original sin: they do the United States' job too well. . . . The possibility for the United States of withdrawing part of their men and tactical nuclear weapons, while claiming that NATO's defence potential is reinforced by the deployment of new weapons, corresponds exactly with their needs. . . . Certain Europeans, then, feel impelled to denounce vigorously the element of wishful thinking in conceptions of PGM which essentially constitute a useful alibi for United States disengagement' (op. cit. in note 59, p. 8).

[59] Alain Enthoven, 'US Forces in Europe. How Many? Doing What?', Foreign Affairs, April 1975, pp. 513–32.

PGM with low-yield nuclear warheads for battle-field and interdiction missions.[60]

In fact, if any overall tendency is discernible in connection with technological developments, it is the blurring of traditional distinctions between conventional, theatre nuclear and strategic nuclear weapons. While low-yield nuclear weapons can carry out tasks now assigned to conventional munitions, missiles with specialized conventional warheads could in the future be used against targets now only vulnerable to nuclear attack, including some in the Soviet Union. This development may be difficult to resist, and for those who emphasize the coupling of European defence to the United States in a 'seamless web' of deterrence it is an important development. In a situation where NATO maintains the capability to respond to military threats with a range of overlapping nuclear and non-nuclear responses, including the use of both on the Soviet homeland, a strategy of maximum flexibility increases the ambiguity facing Soviet planners and at the same time reassures Europeans of the American commitment. A seamless web of deterrence appears to be in the American interest because it relies on conventional defences and emphasizes the careful and controlled use of nuclear weapons. For Europeans, the ever-present threat of escalation – with its impact on overall deterrence – is an appealing factor.

Closer examination of the concept reveals some problems, however. In seeming to be all things to all men it masks, but does not resolve, the clash of interests within the Alliance over when and how to use nuclear weapons. True,

the threshold between nuclear and non-nuclear weapons is breaking down where the destruction of point targets is concerned, but the threshold still has powerful political and psychological significance. In a time of strategic parity, and when fixed-site ICBM are increasingly vulnerable to attack, the danger of escalation to full-scale nuclear war is implicit in the link between nuclear use in the European theatre and the American strategic nuclear arsenal. In this situation, there is growing value in de-emphasizing the role of nuclear weapons in the overall spectrum of deterrence. For some Europeans, however, too much emphasis on conventional weapons suggests a prolonged and destructive conflict – but emphasizing a new generation of low-yield nuclear weapons suggests an even worse fate: a 'limited' nuclear war in Europe with both super-powers as sanctuaries. Nonetheless, this does not mean that Europe must accept the least distasteful alternative and resign itself to bolstering NATO's conventional capabilities at the expense of nuclear threats. A nuclear 'coupling' mechanism is still required to maintain Alliance cohesion and win European acquiescence to a conscious American effort to decrease NATO dependence on the threat of escalation to deter war. At present that mechanism seems to exist, in the form of the doctrine of flexible strategic options; but, though American officials have emphasized that the selective targeting concept is a response to the need to maintain a strategic nuclear link to Europe, the circumstances that might cause selective use of American strategic weapons to defend Europe have yet to be clearly spelled out.

At this point it is interesting to explore the impact of technological developments on the concept of the 'threshold'. In the past, strikes on strategic targets have been thought to require nuclear weapons, the vast majority of which are under the direct control of the United States. But, since long-range, terminally-guided ballistic or cruise missiles with conventional warheads could now strike at certain targets deep in eastern Europe and within the Soviet Union, it would be possible for NATO to cross 'territorial' thresholds without running the risks attached to nuclear use. The spectre of a war (nuclear or non-nuclear) limited to Europe might thus be escaped in a manner which could be acceptable to the whole Alliance.

---

[60] The tendency has been to view the new weapons technologies as a way of substituting conventional munitions for nuclear weapons in various roles, but this is not inherent in their exploitation. First, the longer-range terminally-guided delivery vehicles now under development (cruise missiles, the improved versions of the *Lance* and *Pershing*) will be dual-capable. Second, technologies for improving the effects of low-yield nuclear munitions are available. Thus, while new conventional munitions offer numerous weapons-tailoring possibilities, the design of low-yield nuclear warheads also promises new targeting options. Minimum-residual-radiation explosives can reduce collateral damage when this is important, while enhanced-radiation weapons are efficient in the anti-personnel role. The use of the latter to incapacitate opposing forces is discussed by S. T. Cohen, *On the Stringency of Dosage Criteria for Battlefield Nuclear Operations* (Santa Monica, Calif.: The Rand Corporation, P-5332, January 1975.)

The possibility of strategic conventional warfare, however, again raises the question of who controls the course of escalation. It was suggested earlier that acquiring a new generation of weapons could make weaker Alliance members more dependent on larger arms suppliers. But the United States' ability to control the escalation process could decline when other NATO states were able to launch punitive strikes on Eastern Europe, or even the Soviet homeland, with accurate conventional munitions. The profusion of different threshold-crossing possibilities (in terms of state boundaries, civil or military targets attacked with nuclear or non-nuclear weapons launched by major or minor powers) creates numerous instruments of escalation, some of which European states will possess. The price the United States may have to pay for raising the nuclear threshold might therefore be the lowering inhibitions on engaging in strategic conventional warfare against the Soviet Union. This prospect raises its own special questions. Would, or could, the Soviet Union differentiate between nuclear or non-nuclear strikes on her homeland? Would deployment of this new class of systems encourage a neglect of other conventional capabilities, thus necessitating use of these systems in the event of war and increasing the risk of escalation? Would American efforts to limit European acquisition of strategic conventional capabilities lead to increased fears of disengagement and super-power bilateralism?

East–West Arms Control
These questions provide some indication of the daunting problems that new technologies provide for arms-control institutions. In general, the erosion of traditional distinctions between weapons threatens the carefully constructed political and geographical definitions underpinning the Strategic Arms Limitation Talks (SALT) and the negotiations on Mutual and Balanced Force Reductions in central Europe (MBFR). At SALT, new long-range delivery systems raise questions similar to those posed by non-super-power nuclear forces and by so-called 'forward based systems', but the capabilities offered by emerging technologies suggest that each side will attach much greater importance to constraining new options available to the other. This has already been shown in the controversy and deadlock over the American cruise-missile programme. When fitted with a map-matching guidance unit, nuclear-armed cruise missiles now under development will be capable of attacking strategic targets in the Soviet Union. If, as projected, the system is deployed in the early 1980s in air- and sea-launched versions (the latter delivered from surface vessels or the torpedo tubes of submarines), failure to include these systems under launcher ceilings agreed at SALT would defeat one of the main objects of the current phase of the negotiations: eliminating uncertainty over the future size and character of adversary forces. But if both sides agree in principle to constrain cruise missile deployment, major difficulties must be faced: constraining cruise missiles would limit their use not only in strategic nuclear roles but also for tactical missions. Moreover, an agreement that singled out specific modes of cruise missile for limitation would also discriminate against one or other super-power, because of asymmetries in air-defence capabilities, location of targets (on coasts or inland), and access to ocean operating areas.

Possibly more intractable is the difficulty of monitoring an agreement that includes cruise missiles; their potential for deployment in large numbers aboard stand-off aircraft or attack submarines would seem to make adequate verification virtually impossible.[61] But to understand the cruise missile dilemma purely in terms of verification requirements is to underestimate the nature of the problem. A central assumption underlying the SALT exercise of launcher limitations is that strategic delivery vehicles can be identified and counted. The adaptability of cruise missiles to both strategic and tactical roles makes it increasingly difficult to differentiate between weapons suited for one or the other. Therefore, while it is still possible to distinguish strategic from tactical missions, it may soon no

---

[61] The first round of SALT accords do not constrain delivery vehicle numbers *per se* but launcher numbers – ICBM silos and SLBM tubes. To limit cruise missile launchers (rather than the missiles themselves) it might be necessary to constrain the numerous delivery vehicles systems capable of being fitted with these systems (an approach that the Soviet Union apparently favours) by counting certain launchers fitted with cruise missiles as MIRV systems while prohibiting their deployment on other potential launchers. If the missiles themselves are made subject to control means must be found to monitor their deployment aboard submarines, aircraft, surface vessels and with land units.

longer be possible to identify categories of weapons specific to one or the other.[62]

It is important, then to examine the arms-control problems that might accompany the deployment of modern cruise missiles in tactical and strategic roles in and around the European theatre? How would including cruise missiles in a new SALT package affect European and American options, and what impact would this have on the political framework of the Alliance? For tactical roles, the prototype cruise missiles under development by the United States will be compatible with mobile ground launchers used for existing tactical ballistic missiles and will be able to deliver a variety of conventional payloads over ranges exceeding 500 miles. While cruise missiles with these characteristics would give NATO forces a number of new targeting options in the European theatre, the ability to convert them into longer-range nuclear-armed delivery vehicles for use against targets in the Soviet Union gives the Soviet leadership a strong incentive to limit their deployment in Europe. Consequently, if strategic versions of the cruise missile are included under a SALT launcher ceiling, the Soviet Union is likely to insist on non-circumvention rules that would constrain the deployment of tactical versions – a step that could prejudge efforts to reorganize theatre defences on the basis of the new capabilities the systems offer. Even more adverse to European interests would be non-transfer language prohibiting the export of cruise missile technology, modelled on Article IX of the 1972 ABM Treaty. Because of the problems of definition and verification that cruise missiles pose, the two sides may finally decide not to include them in a new 'interim' agreement on offensive forces. But leaving them out of a SALT II regime would only serve to reinforce pressures to make forward-based systems – missiles and aircraft – a central concern in the third phase of the negotiations. Even deferring the cruise missile issue to a later round of talks could have a divisive impact on the Alliance if the Soviet Union adopted an attitude towards these systems similar to the American position on mobile ICBM after 1972, announcing that their deployment by Europeans would be 'incompatible' with a new arms-control regime.[63]

The potential deployment of a new family of nuclear weapons and delivery vehicles in central Europe poses similar problems at the MBFR negotiations. In December 1975 the NATO participants at the Vienna talks proposed that Western nuclear weapons, missiles and aircraft be included in a first-phase American–Soviet reduction package that would also entail a reduction of Soviet tank forces (the most worrying component of Soviet forces in central Europe). It is not clear, however, how this proposal will affect possible American efforts to modernize existing nuclear stockpiles, or whether American or European attempts to explore new delivery vehicle options – such as precision-guided ballistic missiles, cruise missiles or RPV – might be pre-empted by the inclusion of aircraft and missiles in an overall force-reduction package. MBFR non-circumvention language could also narrow future Western (and especially European) deployment choices, while a SALT agreement that severely constrained new delivery vehicles could prejudge MBFR.

In general, the introduction of new weapons technologies creates uncertainties that make the calculus of arms control more difficult. Attempts at MBFR to arrive at mixed packages that reflect NATO and Warsaw Pact asymmetries will be further frustrated by the changing vulnerability of systems. In a period when, for instance, tanks are seen to be increasingly vulnerable to a new generation of anti-tank weapons, calculating arms-control trade-offs becomes, for all practical purposes, impossible. If tanks stand to be more vulnerable, and so less valuable for purposes of bargaining, in 1985 than at present, an agreement reflecting a high priority given to limiting Warsaw Pact tank numbers could be bought too dearly in terms of the present and future NATO capabilities traded away to reach the accord. The difficulties of reaching agreements will also be

---

[62] This is not a new problem. In the area of anti-submarine warfare (ASW), it has long been recognized that many naval systems can undertake both 'strategic' and 'tactical' ASW operations, and this flexibility has probably been the most important conceptual obstacle to efforts in SALT to constrain ASW activities.

[63] The cruise missile might provide a relatively inexpensive means of enhancing the size and survivability of British and French sea-based strategic forces. If cruise missiles were deployed aboard British nuclear attack submarines, the size of the British strategic submarine force would grow to 15 (4 *Polaris*-type and 11 attack submarines) by the end of the decade.

exacerbated by uncertainties within defence establishments over the future shape of forces. Many people maintain that the tactical aircraft has become too risky and expensive a means of delivering munitions against heavily defended targets – but, until the problems of replacing manned aircraft with alternative systems are better understood, manned aircraft are likely to remain the primary instrument for deep interdiction. Thus, until firm decisions over alternative delivery systems can be made, there will be strong and understandable resistance to closing options through arms-control agreements.

In these circumstances future East-West arms control might be better directed towards exchanging information and working out understandings, rather than constraining numbers and types of delivery vehicles. In view of the uncertainties that new weapons will introduce to the tasks of discerning capabilities and intentions, emphasis on confidence-building measures might be arms control's most appropriate task at present; prior notification of the development of new systems, or of the redeployment and reorganization of forces, and the clarification of ambiguous policies would be of special value in a period when uncertainty over the military impact of new weapons will encourage a natural hesitancy to enter into new agreements. But in pursuing this approach the central question is whether the Eastern nations, and the Soviet Union in particular, are prepared to engage in arms control that aims to achieve greater transparency in military behaviour; the Soviet record to date does not give grounds for optimism.

In the longer run, completely new organizational arrangements for arms control may offer greater promise. The blurring of distinctions between strategic and tactical weapons has made it increasingly difficult to conduct SALT negotiations on a bilateral basis over weapons that not only defy categorization but also have important implications for allies not participating in the talks. And it has become just as difficult to conceive of multilateral negotiations on a regional basis at MBFR that could have important implications for the separate super-power strategic relationship. What technology seems to be doing is to force us to recognize two long-standing characteristics that hamper East–West arms control:

(1) The political independence of West Europeans which, if directed towards exploiting a new generation of missiles, could undermine the effectiveness of a strictly American–Soviet arms-control dialogue.
(2) The geographical character of the East–West relationship which, on the one hand, allows unrestricted Soviet deployment of weapons not treated at SALT but which can carry out strategic missions against European targets and, on the other, leads NATO states to deploy in Europe weapons for tactical roles which, in Soviet eyes, pose a threat to the Soviet homeland.

Arms-control efforts must recognize that the American–Soviet strategic relationship and the East–West military balance in Europe are becoming more closely intertwined. If arms control is to retain its existing focus on controlling numbers of aircraft, missiles and warheads, it may ultimately be necessary to incorporate the SALT and MBFR processes into a comprehensive East–West negotiating framework that would be better able to link the complicated relationships now emerging.

Arms Transfers and Regional Stability
While most of the new systems are being developed for the European theatre, the incidence of conflict in the post-war era suggests that the new technology will have a more important impact on military and political developments in other regions. But its implications for conflict in the Third World (with the possible exception of the Middle East) has received insufficient attention. Analysis has yet to go below the level of intriguing generalization. First, it has been suggested that, because the defensive capacities of weaker states are likely to be improved with modern anti-tank and air-defence systems, regional stability in the Middle East, Southern Africa or South-east Asia will be strengthened. Second, the dependence of weaker powers on suppliers of advanced equipment is said to enable supplier states to play a more active 'crisis management' role in the event of conflict. Third, the capabilities promised by a new class of conventional systems could serve to dampen the appetites of some states for nuclear weapons.

Because of the tremendous geopolitical differences between regions, these propositions must be examined with even greater care than those offered for the European situation. The sophistication of military equipment, the environments in which it could be used and the technical competence with which it may be maintained vary considerably. With this fact in mind, some questionable assumptions can be identified underlying the three propositions outlined above.

One is that the proliferation of advanced weapons will proceed in a balanced manner, and that states in a given region will achieve roughly symmetrical capabilities. This appears highly unlikely; nations with greater resources and technical sophistication will be the first to acquire the new weapons. An obvious example is Iran, which has deployed new aircraft from the United States before their deployment in Europe. While those of her neighbours that have the oil revenues to spend are also moving quickly to acquire advanced weapons, there is nonetheless little question that Iran has achieved a position of military predominance in the Gulf.

But, if advanced weapons are asymmetrically acquired, this calls into question the stabilizing influence ascribed to the new systems. 'Stability', in the minds of some observers, flows from the ability of even the weakest state to resist the aggressive designs of more powerful neighbours. As defined by regional powers like Iran, South Africa, Brazil or India, however, 'stability' could come to mean the unchallenged military capability to control events within a given sphere of influence.

The case of Iran also casts doubt on the assumption that states supplying advanced weapons, can use the 'dependence' of the recipients to control the transfer process in order to diminish chances of conflict. This problem has several aspects. First of all it has been argued that the United States has been able to guard against a pre-emptive South Korean invasion of the North by denying Seoul a large, modern tactical air force; in the same manner, it is said, she can exert some pressure on Israeli policy by manipulating arms supplies. However, the reverse is also true: the increasing reliance of major powers on clients for the performance of vital functions (in Iran's case, policing oil transit routes in the Gulf and the Indian Ocean) may place a weapons supplier in a dependent position.

Probably more important in terms of the spread of technology is the emergence of a 'buyer's market' for advanced armament, the result of a growing number of alternative suppliers and the emergence of newly-affluent recipients. This phenomenon has two primary effects: it enables buyers to exercise greater freedom of choice, and it allows richer states to become 'middlemen' in the arms-transfer process, either by supplying funds for the purchase of weapons or by actually supplying weapons in time of war. In a buyer's market, the most conspicuous aspect of Iranian military expansion has been the purchase of over 1,500 tanks and 400 combat aircraft. Saudi Arabia, meanwhile, is able to influence Middle East events by making similar systems available to front-line Arab states. Pakistan, on the other hand is a different sort of middleman, offering aircraft crews and service units to Gulf states that could not otherwise operate advanced aircraft – perhaps in return for some claim on these systems in the event of war in the Subcontinent. It is in such ways that the possibility of controlling arms transfers to the Third World in a strategy of bolstering stability by strengthening the defence capabilities of weaker states is frustrated.

Major arms suppliers bear some responsibility for this predicament. Not only is selling strike aircraft and armoured vehicles often more lucrative than selling other equipment, but military assistance programmes have often neglected to take into account the special requirements imposed by different regional environments. The tendency of major powers to act as 'models' for smaller states does not extend only to the purchase of supersonic jets for impressive fly-bys on national holidays, but also affects the training and equipping of forces. Thus, despite Geoffrey Kemp's observation that 'in general, simple, rugged arms that are extremely mobile provide many of the Third World forces with more effective military capabilities . . . than high-firepower weapons', the major arms suppliers have generally persisted in exporting doctrine and equipment that are inappropriate to new military situations.[64] To the extent that

---

[64] Geoffrey Kemp, 'Arms Traffic and Third World Conflicts', *International Conciliation*, March 1970, p. 33. A classic example of the tendency to apply inappropriate force-design criteria to new environments was the modelling of South Vietnamese counter-insurgency forces on the American infantry division during the Indochina conflict. See Guy Pauker, et. al., *op. cit.* in note 28, pp. 28–31.

NATO countries, particularly the United States, fail to adapt their own tactics and organization to the performance of the new weapons, therefore, recipients of Western military assistance are likely to have similar problems. The application of inappropriate criteria to the design of Third-World forces is likely to be a greater problem in a post-Vietnam period, when the United States is once again preoccupied primarily with problems of European defence.

Closely connected to the problem of the asymmetrical introduction of inappropriate military technologies into the Third World is the question of intervention and crisis management. Even if the new technologies are exploited in a manner that bolsters the defensibility of weaker states, this will still have ambiguous consequences. While weaker states might be better able to deter and defend against the encroachments of neighbours, the same states' ability to use advanced weapons to control, or at least disrupt, activity in areas of strategic importance could have an unforeseen impact on major supplier states' ability to secure their own interests.[65] This might be especially worrisome for maritime nations, where littoral states and states athwart straits might use submarines, mines and anti-shipping missiles to seal strategic maritime choke points. The capacity of a small state with a relatively sophisticated force of missile-equipped ships or submarines to disrupt traffic in such congested areas as the Mediterranean, the North Sea or the Sea of Japan would be considerable.[66]

Another dimension of the same problem is the possibility that sub-national groups will exploit the capabilities of the new weapons. Revolutionary or secessionist movements armed with man-portable air-defence and anti-tank missiles, will be better able to resist attacks by government forces, and weaker governments will find it harder to discourage the use of their territory as a sanctuary for guerilla forces fighting in adjacent states. Other states, lacking the capability to mount a conventional armed attack, might be tempted to equip terrorist forces with advanced weapons for use against an adversary in a strategy aimed at sabotage and provoking public alarm.[67]

The overall impact of the new weapons on relations between stronger and weaker states is difficult to predict with certainty. The factors mentioned above imply greater freedom of action by weaker states, but their dependence on technological transfer also suggests greater political influence for supplier states. This is particularly true in actual conflicts, when the expenditure of munitions accelerates, and governments will then be forced to depend on supplier states for rapid resupply. If an original supplier refuses to resupply, the existence of a 'buyers' market' in arms is not in practice very relevant to a client actually involved in a conflict, because of the difficulties of quickly and efficiently integrating unfamiliar systems into existing force structures. And even when there is no conflict in progress, forces already trained and organized to use one supplier's systems face adaptation problems when attempting to switch to weapons from another source.[68]

In the long run, however, indigenous production of advanced weapons would reduce what political influence arms suppliers can exert over recipients. This is the primary motive behind, for instance, Israel's attempt to gain greater self-sufficiency in armaments production—a policy that has led to the development and production of anti-ship and surface-to-surface tactical missiles, the *Kfir* supersonic strike aircraft, and (reportedly) a warhead for the American-supplied *Lance* missile. Israel, with her well-established research and industrial base, is clearly better placed than other states to become more self-sufficient in armaments production, but countries such as India have demonstrated a growing capability for indigenous development

---

[65] The clearest illustration of this dilemma is the American military assistance programme to Saudi Arabia, which is a two-edged weapon: American weapons will improve the capability of Saudi forces to resist military intervention from the outside, including American intervention to secure oil supplies in the event of a boycott that threatened the West with energy 'strangulation'.

[66] This concern reportedly weighed heavy in the British decision in May 1975 not to conclude an arms agreement with Libya which included the sale of six light submarines.

[67] Brian Jenkins calls this 'surrogate warfare', see 'International Terrorism: A Balance Sheet', *Survival*, July/August 1975, p. 164.

[68] There has been much discussion of the possible use American arms aid to reduce Egypt's military reliance on Soviet equipment, but it has been estimated that for American weapons to supplant Soviet equipment in Egyptian armed forces would take 10 years and require at least $7 billion in military aid (*New York Times*, 22 October 1975, p. 3).

and procurement. Efforts by less advanced states to pool financial and technical resources in multi-national production enterprises could speed the appearance of other indigenous arms capabilities: in 1975 Iran, Pakistan and Turkey announced a joint arms-financing and production arrangement, and in February 1976 Egypt, Saudi Arabia, Qatar and the United Arab Emirates agreed (on paper, at least) to co-ordinate their military investment and military production.

Another question raised by the availability of a new generation of conventional arms concerns is the spread of nuclear technology. Will the proliferation of these systems blunt or reinforce the drive for nuclear weapons? To the extent that the new technologies will give states a greater sense of security, it might be argued that they will serve to lessen the incentive of going nuclear. For example, it has been maintained that American assistance in strengthening the Israeli capacity to respond to military threats with conventional forces reduces the urgency of exploring a nuclear option; similar arguments have been made in reference to South Korea. But any attempt to generalize from these cases must take account of the many different considerations that lead states to acquire nuclear weapons. Israel faces a clear and immediate military threat, and her weapons acquisition policies are directed towards countering it. Other governments may be tempted to acquire nuclear weapons (or at least demonstrate a nuclear capability) in order to enhance national status or in response to domestic political considerations – factors which appear to underlie the Indian action of May 1974 – and in such cases the availability of improved conventional weapons would have little impact on the nuclear decision. Even in the case of Israel, the acquisition of advanced conventional equipment might not influence the course of the nuclear debate, for a nuclear capability might be seen as attractive not for war-fighting purposes but as an instrument for threatening escalation of a conflict so that it might ultimately involve the use of nuclear weapons by the super-powers. If the super-powers also foresaw such a possibility Israel might view the 'secret' acquisition of nuclear weapons as a way to obtain powerful leverage over them in time of crisis.

A special case is the possibility that the United States might be tempted to view the supply of improved conventional arms to allies as a means of reducing the American nuclear commitment; as we have seen in the case of Europe, reductions in TNW numbers and in troop levels could be justified by better conventional capabilities. In Europe such a strategy would be likely to spur a re-examination of European nuclear options, but severe domestic and international inhibitions, would act to prevent West Germany and other non-nuclear states from acquiring nuclear weapons. In East Asia, these inhibitions are not nearly as strong. If emphasis on a new generation of conventional systems were seen as substantially reducing the possibility of American nuclear involvement in the defence of South Korea, the pressure on Seoul to acquire nuclear weapons would be considerable.[69] A declining American nuclear presence in East Asia generally would likewise influence nuclear decisions taken in Taiwan and Japan.

Finally, one must recognize that the proliferation of certain of the new technologies will enable aspirant nuclear states to acquire more credible nuclear-capable delivery systems. Though supplied for conventional ordnance delivery, at least 13 different types of nuclear-capable aircraft are now deployed among 14 nuclear 'threshold states'.[70] However, despite their sophistication, aircraft do have certain operational drawbacks – mainly restricted combat range and questionable penetration capabilities against some potential adversaries. For most of the aspirants, acquiring long-range bombers and land- or sea-based ballistic missile systems is ruled out in the near future by the financial and technical resources required (India, for instance, would be hard-pressed to acquire a credible ballistic missile capability against populous targets in eastern China), but new long-range delivery systems now on the horizon – cruise missiles and RPV – may offer an attractive and far cheaper solution to the delivery vehicle problem.

---

[69] President Park has been quoted as saying, 'If the US umbrella were to be removed, we would have to start developing our nuclear capability to save ourselves' (*International Herald Tribune*, 13 June 1975, p. 6).
[70] See *Strategic Survey 1974* (London: IISS, 1975), Table 6, p. 38.

V. CONCLUSIONS

It is always tempting to see technology as the driving force behind military doctrine and organization. Thus there is a sense of inevitability in the arguments of the proponents of new technologies: on the tactical level, the new systems will favour the defence; organizationally, technology will force a reordering of missions and a restructuring of forces; in budgetary terms, that new weapons will save money; and on the strategic level, the technology will raise the nuclear threshold, dampen incentives for escalation and open up new arms-control opportunities. Military technology can indeed be shown to have had decisive implications for the conduct of war and political arrangements over time, but for the next decade the impact of new developments will depend not only on the responsiveness of governments, but also on the resolution of critical operational, organizational and political questions. If the experience of the last thirty years is relevant, these questions will not be quickly answered. It is a cliché that it is dangerous to deploy a new weapon without considering the long-term effects of its deployment – but equally dangerous is the tendency to view technological developments as instant solutions to long-standing dilemmas. Before far-reaching choices are made, several factors must be taken into account.

Performance Constraints
A system's capability to perform a desired role must be tested in a variety of operational situations. Present-generation battlefield PGM will not perform efficiently in adverse conditions, and their more versatile successors will be significantly more expensive. The management of large numbers of RPV presents significant command-and-control problems, and precision-guided and conventionally-armed long-range missiles have yet to be proved either operationally feasible or cost-effective. Target acquisition in fluid conditions remains a problem. In the short run, tanks, aircraft and surface vessels seem to be more vulnerable, but longer-term developments – acquisition of improved armour or exploitation of countermeasures – could nullify the advantages of small precision-guided weapons. If these issues are resolved there will still be the problem of resource constraints, which will necessitate difficult choices among competing technologies. For PGM, for instance, it will be necessary to decide between more expensive, high-accuracy systems and less costly projectiles of moderate accuracy. If in the future, numerical asymmetries will be even more important than at present, the outcome of these choices could be crucial.

Tactical Questions
Can small, dispersed land units equipped with anti-tank weapons provide a convincing capability against sustained combined-arms operations? Can air superiority over the battlefield be maintained with battlefield SAM and less sophisticated aircraft? Can a new generation of missiles supplant manned aircraft in a substantial number of existing roles? Crucial to the answers to these questions are not only performance factors and the counter-measures that a potential adversary might employ but the pattern of adaptation that the deployment of new weapons might provoke in an adversary. While that adaptation could itself be an indirect benefit derived from the new weapons, it could also constitute a new and unfamiliar threat. By making 'short-war' strategies less likely to succeed in land war, deployment of a new generation of conventional weapons could raise the threshold of deterrence against resort to force. But the deployment of such weapons, whether conventional or nuclear, could be a more prolonged and destructive conventional conflict if deterrence fails. There are special questions to resolve about the emerging class of strategic conventional weapons: by enabling states to carry out non-nuclear punitive strikes, cruise missiles and RPV may reduce the temptation to resort to nuclear use, but if their procurement accompanies a decline in other conventional capabilities their overall impact could be to provide new incentives for escalation.

Mission Priorities
If the new weapons are to have the desired impact, it is clear that some modification of existing service roles and missions will have to be undertaken. But the extent to which mission assignments will actually be reordered will depend on the willingness of service bureaucracies to explore the operational and tactical problems

75

outlined above. If the new technologies are introduced without a reassessment of existing missions and force mixes, the cost savings advertised by some will not result. Efforts to restructure military forces must cope with the threat posed by bureaucracies with vested interests in maintaining existing missions and equipment, the danger here being not that technological developments will be ignored but that they will be to perform tasks that technology is making both more difficult and less important.

Political Consequences
The key to discerning the impact of new weapons on political relationships seems to lie in the observation that discord and conflict stem from differing interests and conflicting ambitions – not from the existence or non-existence of weapons. In Europe the new technologies could bolster the capability of Western states to respond to a Soviet conventional threat. But this course of action, on the face of it desirable, could precipitate a crisis in NATO strategy if accompanied by a real or perceived American drift towards nuclear disengagement. The new technologies will not by themselves bridge the intra-Alliance gap between notions of the nature of deterrence and defence. Given the differing attitudes towards the threat of and actual escalation to nuclear use, bridging this gap may be an insoluble problem, but the coming debate on the new weapons within NATO should provide a useful reminder that the solution to political problems does not lie in technological developments.

The same is true in other regions. New developments are more likely to exacerbate the problem of political choice, rather than ameliorate it. This seems particularly true for the two super-powers, since new technologies could threaten their ability to manage events both within their respective spheres of influence and in other regions. In the post-war era, military technology reinforced political and economic tendencies towards a bipolar and stratified system of international power; but the character of new weapons technology seems to be reinforcing tendencies towards a more fragmented and pluralistic international system. While the super-powers' military strength continues to grow, the availability of new capabilities to smaller and middle-level powers could mean that American and Soviet capacity to project their military strength will decline in relative terms. In the West, European acquisition of a new class of long-range missiles could reduce American control over the escalation process and undermine the bilateral character of the East–West dialogue over strategic arms. While it is very difficult to envisage similar tendencies in the East, new weapons technologies could make non-alignment and neutrality a more viable option for European states. In addition, the global spread of both older and more advanced military technologies will increase the capacity of states to resist outside intervention. This, in turn, will not only increase the major powers' difficulties in projecting force into distant areas but also will facilitate the emergence of stronger and more autonomous regional power centres.

Arms Control
Attempts to explore the impact of new weapon developments on arms-control efforts reveal a certain schizophrenia. While developments like the American cruise missile are seen as creating new problems for limiting strategic nuclear arms at SALT, the same systems, when used for theatre tasks in Europe, could strengthen Alliance capabilities. Even when deployed in a strategic mode aboard a stand-off bomber, cruise missiles have been seen as an effective, but far less expensive, alternative to manned bombers designed to penetrate enemy defences to reach their targets. Thus, as the cruise missile case suggests, even the most dedicated advocates of arms control must recognize that new technologies may offer attractive solutions to the classical arms-control goals of bolstering deterrence, minimizing damage if deterrence fails, and saving money. But focusing only on the specific problems and prospects that new technologies introduce into existing negotiations misses the deeper consequences. New nuclear and non-nuclear targeting options will create new threshold-crossing possibilities which will erode the bilateralism of SALT and the artificial geographic limitations of MBFR. As the two sets of negotiations struggle with the growing problem of technical and political overlap, the versatility of new systems will increasingly defeat efforts to define self-evident categories of delivery vehicles in terms of mission, deployment or technical

characteristics, and hence to reach mutually acceptable formulas for limiting their numbers. If this is so, the military and political terms of reference underpinning existing arms-control institutions may be on their way to obsolescence. To be effective against a background of the vertical and horizontal proliferation of new technologies in the 1980s, arms control may have to shift away from controlling delivery vehicles numbers towards mutual understandings about the deployment and operational use of the new systems.

3 Precision-guided Weapons
JAMES DIGBY

INTRODUCTION

Ever since military men began shooting things at enemies, most shots have missed or been ineffective. The remarkable thing that has happened over the past few years is that new weapons have been developed which can hit with most of their shots, usually effectively. While much of the theory of these weapons is not new, it is only since laser-guided bombs were used in Vietnam that military planners have generally agreed that they were economically and operationally feasible. The early name 'smart bomb' was dropped, and a larger class of 'precision-guided munitions' (PGM) became official. Usually this term simply refers to a bomb or missile that is guided during its terminal phase, thus including many anti-tank weapons and air-defence missiles. It was to these that attention shifted after their widespread use during the October war of 1973, especially by Arab forces.

In the first part of this Paper I shall say enough about the mechanics of these weapons to give the reader a feel for how they work, and provide brief descriptions of some of the more important weapons associated with non-nuclear land combat. However, the brief treatment here mentions only a fraction of current PGM developments (it is characteristic of the pace of development that dozens of new PGM types in many countries reach the operational testing stage each year). The second part of the Paper focuses on a number of important and so far unresolved implications of the new weapons, and discusses their likely effects on force posture and the conduct of warfare. For example, what is their effect on the usefulness of the advanced tank, the complex fighter-bomber and the big aircraft carrier? What will be the consequences for the organization of land forces and for their tactics? Land forces may need to adopt a kind of molecular posture of many highly mobile and concealable – but powerful – squads, and it may be that a smaller weight of munitions will have to be hauled to the battle area. What are the political consequences? If barrage fire and carpet bombing are not needed there may be less collateral damage to civil populations and the economy, and there are prospects for raising the threshold at which nuclear weapons would be used (the consequences of this are both encouraging and urgently in need of study). On the positive side (from an American point of view) there is a likelihood that the resulting postures will be advantageous to NATO and to American strategies.

Serious consideration of many of these points has only just begun, and the devising of counters to PGM is just gaining momentum. Thus the reader should regard much of what follows as tentative. After all, military analysts foresaw only dimly the implications of some very predictable technologies, and it took the 1972 bombing raids in Vietnam and the October 1973 war in the Middle East to get serious exploitation of PGM under way.

It is beyond the scope of this Paper to say very much about technological trends or to explore adequately countermeasures and counter countermeasures.

I. EXAMPLES OF PGM

This part describes several precision-guided munitions – of different kinds – to give the reader a better idea of the scope of what will be discussed. A pedantic definition of a PGM would run something like: 'A guided munition whose probability of making a direct hit at full range – when unopposed – on a tank, ship, radar, bridge or aircraft (according to its type) is greater than a half.'

Many discussions exclude surface-to-air missiles, principally because they have been quite precise for a long time; what is new in their case is being precise *and* relatively cheap *and* easy to operate. But note the adverb relatively. Many of the new weapons are quite sophisticated and can only be called cheap in relation to earlier guided missiles or to the destructive capacity of other weapons that might be used for similar jobs. However, they are not too expensive to preclude an abundant supply – and the possibility of abundance accounts for much of the significance of PGM.

General awareness that something new had happened was triggered by the performance of the unpowered *Pave Way* laser-guided bombs in Vietnam in 1972. With fairly simple aids, a narrow laser beam from a 'designator' was pointed at the target from an aircraft. A two-part kit attached to Mk 82 500-lb or Mk 84 2,000-lb bombs provided steerable front fins (controlled by a laser receiver) that homed on the energy reflected from the target. In earlier systems the laser beam was aimed at the target from a spotter aircraft, but a later version avoided mix-ups by designating from the F-4 bomb-carrier itself. Excellent accuracies could be obtained, making it possible to destroy in one or two sorties a bridge span that might otherwise have required dozens. Judging from the aerial photographs released by the US Air Force, the accuracy equalled or exceeded the requirements set in 1968: 'CEP no greater than 25ft; guidance reliability at least 80 per cent.'[1]

The Soviet *Sagger* AT-3 wire-guided anti-tank missile saw extensive use in the October 1973 war in the Middle East. Often mounted in sixes or eights under a kind of steel umbrella on the BRDM-2 armoured car, it weighs 11 kg, has a 2·7kg warhead and takes 25 seconds to reach its maximum range of 3,000m (which is long enough to allow the target to seek cover or to distract the guider by taking him under fire). Similar missiles on the NATO side are the American *TOW* and the Franco-German *HOT*, both of which are faster and have semi-automatic tracking features that mean the gunner need only track the target, rather than having to

'fly' the missile into the target. The British *Swingfire* is less automatic, but its guidance unit can be positioned up to 100m away from the launcher. All three have helicopter-mounted versions.[2]

Let us examine in a little more detail how one of these anti-tank missiles works, taking *TOW* BGM-71 for our example. The missile is 15cm in diameter and weights 23·6kg including its protective container, which serves as a launch tube. The container is sealed at the factory, and the missile is used without further checking. The *TOW* launcher is mounted on a tripod which can be used independently of a vehicle, or, very commonly, on a metal column fastened to the M-113 armoured personnel carrier. The gunner sights on the target vehicle through a 13-power monocular telescope; if there is a choice of targets, he selects the one most likely to stay in sight during the time of flight of the missile. When he presses the firing button the rocket motor ignites, leaving a signature significantly less than that of the 106mm recoilless rifle which *TOW* has replaced. As the missile flies towards the target, control wires attached to the launcher unwind from two bobbins, each about 18cm long, mounted in the tail of the missile. A modulated infrared source in the missile tail is then followed by an infrared tracker mounted in the launcher, and steering commands, based on the difference between the angles from launcher to missile and from launcher to target, cause the missile to move its aerodynamic fins, and direct it to fly a course roughly along the line of sight to the target. Meanwhile, a safety and arming device arms the warhead, a 3·6kg shaped-charge device,[3] which detonates on contact.[4]

In training, *TOW* can engage three targets within a 90-degree angle in less than 60 seconds, though some experienced officers doubt if such a rate would usually be possible in combat. The 1974 production version of *TOW* has a range of 3km and a time of flight over that distance of about 15 seconds. An extended-range version

[1] Quoted in P. G. DeLeon, *The Laser-Guided Bomb: Case History of a Development* (Santa Monica, Calif.: Rand Corporation, R-1312-1-PR, May 1974). This report includes a useful bibliography.

[2] Data from *Flight International*, 14 March 1974.
[3] A shaped charge (or hollow charge) has its explosive so shaped as to focus the blast wave generally along the axis of flight. The hot concentrated gases can burn a hole in armour which might withstand a non-directional blast from the same weight of explosive.
[4] *Jane's Weapon Systems, 1973–74* (London: Jane's Yearbooks, 1973).

has been developed with a maximum range of 3·75km. At present the cost of the missile is approximately $3,000, while the infantry-version guidance unit costs about $20,000.[5]

Maverick AGM-65A, an air-launched anti-tank missile developed by the US Air Force, is steered by a stabilized television camera in the nose which feeds video signals into automatic tracking circuits in the missile; these generate guidance signals. Six missiles can be carried on F-4, A-7, and A-10 aircraft. The pilot manoeuvres to get his sight lined up with the target, and when he does so servo controls lock the missile's automatic tracking device on to the same target. When the cross-hairs of the missile's target-seeker frame the target, a lock-on switch is actuated and the missile's contrast-seeker automatically tracks the target. The missile is then launched. At any time after launching, which can take place several miles from the target, the aircraft can leave, or the crew can repeat the procedure and attack other targets. *Maverick* is 30cm in diameter, carries a relatively heavy warhead and has a total weight of 210kg. For the quantity procurement begun in 1974 the unit price was under $10,000. A laser-guided version and an infrared version are under development.[6]

American forces first saw the Soviet *Grail* SA-7 anti-aircraft missile used in action in Vietnam.[7] It can be carried by an infantryman and launched from the shoulder. It has an infrared seeker and a 1·1-kg warhead. Reports on the October war indicate that it was fired in salvo at single Israeli aircraft, 'damaging the jetpipes of many Israeli A-4s but not achieving a very high kill ratio'.[8] However, the Egyptian forces which used the SA-7 probably proved its operability. The SA-7 is similar to the current American *Redeye*, which is being replaced by an improved version called *Stinger*. The British *Blowpipe* is an optically guided missile of similar size, also fired from the shoulder.

---

[5] *Ibid.*, and *TOW Weapon System Status* (Hughes Aircraft Company, December 1973).
[6] *Flight International, op. cit.*, p. A1.
[7] The Soviet Union does not normally give official nicknames to such weapons; *Grail* is the NATO nickname. Journalists in Indochina reported the name as *Strella*, which may have been a phonetic spelling, via Vietnamese, of the Russian *strela* (arrow).
[8] *Flight International, op. cit.*, p. A12.

Standard ARM (anti-radiation missile) AGM-78 has been developed by General Dynamics for use by US Navy and US Air Force attack aircraft against missile radars and other radiating targets. The seeker design is based on that of the *Shrike*, which saw extensive use in Vietnam against surface-to-air missile radars. A signal-intercept unit aboard the aircraft processes received signals and gives the missile homer the necessary data to permit it to lock on. After launch a passive radiation seeker steers the missile;[9] if the hostile radiation source shuts down, stored information is used to guide it on a less accurate trajectory. Anti-radiation missiles, unlike all the PGM described above, are fully operable in darkness and bad visibility and are particularly useful for suppressing such emitting targets as SAM radars. This growing class of weapons is likely to be much more widely used in the future against other emitting targets, such as command and control centres.

Consideration of two hypothetical missiles of somewhat larger size will round off this list of typical PGM. While components for missiles like these are in development, and missiles even have been tested as complete systems, no fully operational weapons like these exist in quantity.

First, there is the possibility of a 100-km range air-launched cruise missile that gets its mid-course guidance from ground-based transmitters, then, for the last 10km, it corrects its course to target by means of an optical area correlator. This is a device whose electronic circuits compare a map-like picture of the terrain below (received by an on-board sensor) with a reference picture taken on a reconnaissance flight – probably from very high altitude. This kind of missile would be well-suited to attacking a fixed target, such as a depot or airbase, and its map-matching terminal guidance would be very hard to jam. It could, with a well-organized operation, be used against sporadically moving targets.

Second, let us consider a possible alternative way of doing the same job. This could be an air-launched remotely piloted vehicle (RPV) of about the same size and range, guided by a pilot who watches a television picture relayed from the RPV and sends his steering and throttle signals by radio. In addition to attacking fixed targets, this missile could be used against moving or

---

[9] A 'passive' seeker receives signals but does not transmit; an 'active' seeker has both transmitter and receiver.

movable targets. The US Navy's *Condor* AGM-53A already has these properties, among others, but refinements in guidance and anti-jamming features have driven its price up to more than $200,000; a simpler design, more like the Advanced Research Project Agency's experimental *Praeire*, a remotely piloted model aircraft of lower performance, may be able to do many of the same jobs.

The foregoing is enough to indicate that the term PGM covers a broad class of guided bombs and missiles. RPV are usually included if they are intended to hit a target (these are sometimes called 'kamikaze RPV'.)

II. IMPLICATIONS OF PGM TWENTY YEARS HENCE

Let us now turn from what PGM are to some speculations on the consequences of having them. I will use a kind of funnel approach, taking the broader and longer-term consequences before coming to the nearer future and to some specifics relating to NATO. By discussing PGM in the context of operations in the 1990s we will not be held back by practical and bureaucratic constraints, as we will be in discussing implications for the earlier period.

The basic point about precision-guided weapons can be stated thus:

Accuracy is no longer a strong function of range, and if a target can be acquired it can usually be hit. For many targets, hitting is equivalent to destroying.

A second point may be equally important:

Precision-guided munitions can now be mass-produced in great quantity; for many the cost per round ranges from around $1,000 to around $10,000. Moreover, many can be operated by average soldiers.

A number of important propositions follow logically from these statements. I will outline seven of them in simple terms before discussing some complications and the degree to which the simple ideas are applicable in the practical world.

Proposition 1. It will become much less desirable to concentrate a great deal of military value in one place or in one vehicle. This will be especially true where a great deal of value can be destroyed by a single warhead. For instance, a combatant would be less likely to want to place a large fraction of his capability at risk by exposing a single transport aircraft, or a single surface vessel in the Mediterranean, and he would probably prefer to have many inexpensive lightly armoured vehicles rather than fewer more expensive tanks. If the attacker has a finite number of PGM, any one of which has a high probability of destroying its target, then it is better to force him to spread them over many targets which are individually of small value.

Proposition 2. With PGM, seeing a target can usually lead to its destruction. Concentrations of vehicles or men – usually easier to see and keep track of than larger numbers of independently moving targets – will be less practical, and concealment will become more important. Smallness and mobility will make hiding easier, and both these qualities are consistent with the thrust of Proposition 1.

However, one must also consider the degree to which concentrations can still be sheltered, or protected by active defences. It is a classic offensive tactic to attempt overwhelming superiority in a narrow sector by concentrating forces there, and the ability to organize and defend such concentrations is an important factor in any assessment of the balance between offence and defence. The availability of tactical nuclear weapons did not in practice result in fully corresponding action to decrease vulnerability perhaps because of uncertainty about whether they would in fact be used. However, there is no question of PGM not being used if fighting takes place, and no tactical planner can any longer afford to ignore their effect on his vulnerabilities.

Proposition 3. Even small units can be very powerful when equipped with PGM or with designators that can call in and guide remotely-launched PGM – and they might carry air-defence weapons as well. In land warfare the natural size of many such independently mobile squads might be three or four men, moving on their feet or in inexpensive vehicles, rather than in expensive tanks.[10]

---

[10] Over the past two years T. F. Burke of The Rand Corporation has developed this and a number of related ideas and discussed them in lectures at the Army War College and other service schools; no published version is available.

Taking the European case, let us suppose that NATO forces were the first to change appreciably towards such numerous, dispersed, concealable, independently mobile front-line units. This would help to reduce vulnerability to both PGM and nuclear weapons, and so would make sense as a unilateral move – so long as the ability of these units to deal with a Warsaw Pact armoured thrust is not seriously degraded. They would also need protection from conventional overrunning attacks by infantry, but their mobility and their ability to call in PGM firepower – especially PGM with area coverage – would help. These would seem to be good reasons, however, for both sides eventually to move increasingly towards a kind of molecular posture for forward units. One can speculate that the forward edge of the battle area[11] would thus become even fuzzier, and, in time, these units would have fewer important targets to designate (the pressures to reach for more distant, rear-echelon targets will be discussed later).

Proposition 4. Where forward units serve as spotters and designators, not all the munitions used need to be hauled all the way to the forward edge of the battle area; some might be ground- or air-launched from tens of kilometres further back. Furthermore, in many types of conflict the higher hit probability of PGM means that, to achieve a given effect on enemy forces, the weight of munitions delivered to the launch point need not be nearly as great as in the past. Nonetheless, the value of munitions used per day may be very large (this will be discussed later). One must consider the changes in both total needs and rate of use before one can understand the implications for the size of support elements and the vulnerability of supply lines.

Proposition 5. A natural consequence of having their high hit probability is that PGM are likely to cause much less collateral damage to civilian populations and economies. In the NATO case, this prospect may in the long term have substantial consequences for West Germany's attitude towards preparations for actual fighting on her territory and the damage it might involve. However, this change of attitude would be much less likely if precision-guided sub-kiloton 'mini-nukes' were under consideration (this, in addition

to the disadvantages of mini-nukes in blurring the firebreak, is mentioned below on p. 88.

Proposition 6. Ground-based anti-aircraft defences will become extremely lethal. The Soviet SA-7 is a step towards a potentially powerful weapon for the air defence of mobile units, and proves the operational feasibility of this class of weapons; however, as already implied, it seems to be under-designed in warhead, range and speed, even against present-generation aircraft. These deficiencies must have become apparent to the Soviet Union during the October war, and correcting them should be a routine matter. In any event, air defences derived from systems like the ZSU-23-4 four-barrel gun, the SA-6 mobile missile and the SA-7 are likely to proliferate over the area occupied by ground troops. The lighter of these classes of weapon may well be added to the mobile squads mentioned above, and the heavier may travel with them, along with the anti-tank weapons. The result may be a shift in methods of protecting ground forces against enemy aircraft: more protection is likely to be provided by ground-based anti-aircraft defences, and less by air-to-air duels and attacks on enemy air bases.[12]

One point in favour of such anti-aircraft defences is the high 'contrast' (in terms of both visibility and energy emission) of an aircraft against a relatively blank sky; this facilitates target acquisition and homing by PGM. Another is that a multi-purpose fighter aircraft costs around $10,000,000 while the weapon to shoot it down costs less than $10,000. These ground defences may not have a kill probability even as high as 5 per cent for the area defended by any one weapon, but flying over many defended areas will be very costly.

Proposition 7. Finally, the properties of these new weapons may well lead to a major revision of the assignment of roles and missions to the different services. It is no longer very important what form of transportation carries a munition to the place where it is launched; it gets its

---

[11] This term is used so as to avoid the implication of shallow depth in 'front line'.

[12] NATO's response to the substantial build-up of Warsaw Pact short-range anti-aircraft weapons has been an emphasis on stand-off air-delivered close-support weapons and on defence suppression. Against aircraft so armed, air-to-air combat is still important. This response – logical, given present NATO aircraft inventories – will in the long run have to compete in cost-effectiveness with close support by RPV, ground- or air-launched from well inside friendly territory.

effectiveness from its warhead and its terminal guidance. This makes it more logical for forces to be organized in terms of the type of target to be attacked. In NATO, for example, the job of dealing with an enemy ship in the Mediterranean has traditionally been a navy job, but if PGM are used it is not immediately apparent whether it is most efficient to employ a ship-launched, air-launched or ground-launched missile, or some combination of these. Organizationally, a task force with no bias against any of the three types would be best equipped to plan the attack. In the long run, an organization specializing in the *task* would be best suited to decide the allocation of money between the various weapon types. Similarly, the job of attacking air bases might be handled by ground-, sea- or air-launched PGM, and it might be more efficient to allocate anti-aircraft defence funds between fighters, sea-based systems and land-based systems from a single common budget.

Some complications

The practical application of PGM will naturally have its full share of complications, and using PGM will be more involved than it appears from the seven propositions above.

To begin with, the technology for accurate guidance that is most fully developed at the moment requires transmission through the atmosphere in or near the visible spectrum (see Appendix I, pp. 91–94, since simple radar guidance is not sufficiently accurate. Many present systems therefore do not work at night, or through smoke, clouds or heavy dust. Systems using long-wave infrared sensors (which will be in use by 1980) will be useful at night and will do fairly well through smoke, dust and haze, but they will be fairly expensive and may be significantly harder to maintain in the field. Nevertheless, for many years the majority of PGM will still require clear daylight.

Another problem is command and control. In past wars, commanders tens of miles behind the front concerned themselves with entire enemy divisions, or, at the very least, battalions. With PGM a division may consist of 500 separately targetable, individually moving objects. The temptation will be to handle this problem from a centralized operations room by means of data-processing technology. (There has been a trend in recent years for higher-echelon commanders to make full use of the profusion of multi-channel communications gear supplied by all-too-willing signal officers. Some senior American officers have called for an 'automated battle-field', and now some Soviet military writers are calling for a 'cybernated battlefield'.[13]) My own judgment, however, is that dealing with precision weapons will require a reversal of this trend. While it will be necessary to draw heavily on advanced data-processing techniques, especially for allocating weapons, I believe much of the solution will be found in delegation of authority and the use of standing procedures, even though the officers doing the detailed weapon control may well also be many kilometres away from the target.

A third complication is what we might call the 'sublimation' problem. If the units near the forward edge of the battle area become too small, too mobile or too well hidden to target, then the natural tendency will be to target depots and other valuable concentrations in the rear-area support structure. Thus, there is likely to be a shift to targets further and further back as longer-range PGM capable of the crucial task of finding such targets are introduced.

In the European context, this shift might find NATO at a relative disadvantage for some years, since it has been the NATO (and especially the American) style to build great depots and rely on a much larger support structure than the Warsaw Pact forces use. Quite apart from any argument for making forward forces less vulnerable, the simple fact is that, as stand-off missiles get better and more practical, action must be taken to reduce the vulnerability of rear-area concentrations – even those several hundred kilometres back and formerly thought safe from any but the most determined air attack. Like several other moves to improve prepared-ness for the use of PGM, this would also make NATO less vulnerable to nuclear attack, and thus help make a nuclear attack less attractive.

A further consequence of shifting attacks to targets further back will be some new attitudes towards sanctuaries. For example, the vulner-ability of NATO's rear-area targets (except atomic-capable aircraft) has seldom been a major subject of concern, but now priorities must be calculated for the protection of *any* concentration

---

[13] See John Erickson, 'Soviet Military Power', in *Strategic Review*, Spring 1973, pp. 71 and 103–6.

of military forces or equipment targetable by stand-off weapons.

The most important complication to the simple picture is added when we consider that countering PGM will take on a very high priority (indeed, work on various kinds of counter-measures is already under way). Concealment and camouflage may work very well against present PGM, and, when they do, an attacker might logically revert to area barrage fire or to area bombing. We must therefore pause before deciding to use cities as defensive strong points. Secondly, the crews of most present PGM are vulnerable (as are airborne platforms) and will be the focus of counter-attacks. Thirdly, new designs of armour may force increases in the size of warheads – which, with shaped charges, can now be quite small.

Let us consider Proposition 1 again. There are complex questions of balance raised by the choice of 'many inexpensive' instead of 'fewer more expensive' vehicles. One has to ask whether the inexpensive vehicles will have the required speed, range and payload. Will the manpower required make the 'many' less desirable? Will only the 'few' be able to mount effective counter-measures devices?

There are problems, too, with the avoidance of concentrations discussed in Proposition 2. Dispersed forces may be inefficient to operate; and can an attacker's 'overwhelming superiority in a narrow sector' not be achieved by calling in offensive PGM from far away, thus concentrating the firepower but not the forces?

In fact, before my seven propositions can move out of the tentative category and become military axioms, there would need to be some force-on-force calculations of a type not carried out to date. While we can marvel at a $3,000 *TOW*-sized PGM being able to kill a $500,000 tank, we really need to calculate how many of these relatively short-range anti-tank weapons would be required on an entire front. At the same time, it would be necessary to compare a system like *TOW* with one where the individual PGM might cost more but be effective over a much wider area – for example, an RPV of 50-km range. One has to think back to the words of General Giulio Douhet, who urged the destruction of enemy bombers in the nest and not on the wing: 'How many guns [in World War I] lay waiting month after month, even years, mouths gaping at the skies on the watch for an attack which never came!' However, whether one talks about anti-aircraft or anti-tank defence, neither one-on-one calculations nor sweeping observations like General Douhet's tell the central story. This would be illuminated by force-on-force analysis, followed by considerations of how the new-style forces might affect plans and intentions as well as the interaction of forces in actual combat.

These thoughts lead naturally to some exceedingly important questions about Soviet strategy *vis-à-vis* the West and about NATO's ability to defend and deter attack. Is the present design of the Soviet Army appropriate to the task of an anti-NATO offensive? Would prudent Soviet military judgment call for a less tank-heavy posture – a ponderous move – before certifying readiness to attack?[14] These are among the most important questions to bear in mind while considering the near-term factors and those specific to NATO, which are discussed in the next two sections.

III. IMPLICATIONS UP TO 1980

It seemed useful to discuss warfare of the 1990s first, to give a sense of direction. But some quite important changes are already upon us. In this section we examine changes that will be important over the next five years, changes that are already affecting force postures and procurement decisions.

The following are some weapons developments with important consequences for our present consideration:

(1) Weapons which, though small, have effective anti-armour warheads – like the Soviet-built RPG-7 small unguided rockets used to good effect by the Egyptians in October 1973.

(2) Anti-aircraft weapons, operated by individuals or small crews, ready to use immediately after movement and cheap enough to be

14 See Col. Edward B. Atkeson's article with the intriguing title: 'Is the Soviet Army Obsolete?', *Army*, May 1974, pp. 9 ff. (with a critical note by C. G. Jacobsen).

available in large numbers. These include several hand-held missiles, such as the Soviet SA-7 already mentioned. In the October war these weapons, along with Soviet-supplied SA-6 missiles and ZSU-23-4 guns, provided such good protection that Egyptian troops could advance without friendly air cover.

(3) Helicopters. The war in Vietnam showed the value of helicopters where opposing defences permitted. Units could be moved to difficult places without being isolated, and light payloads could be delivered tens of miles with little regard to intervening terrain. For our present purposes the point of special interest is that precision weapons are light – and so pack a great deal of capability into a helicopter-sized payload.

(4) Precision weapons for use against surface targets. These are available in great quantity: the Soviet Union supplied hundreds of *Saggers* to her Arab allies, and the United States budgeted for 30,000 *TOW* missiles and 6,000 *Maverick* in FY 1975.[15]

The important consequence of these weapons is that – setting nuclear war aside – the military balance between large-scale forces is likely to be dominated through the 1970s by a new war of numbers. The \$100-million cruiser, \$500-thousand tank and \$10-million fighter will be challenged by the proliferation of less expensive weapons. Most are light enough to be moved easily, and many operate with almost no set-up time. There will be competition to field quantities of these relatively cheap weapons and to design them so that only modest skills are needed to operate them.[16]

However, moves to counter these abundant weapons are under way and are receiving a high priority. Spaced or array armour has been developed to counter shaped-charge warheads, in effect forcing up the weight of warhead needed to knock out a tank and, as a result, reducing the mobility of the PGM system; the United States has already designed an aircraft,

the A-10, to stand up to ZSU-23-4 rounds; countermeasures against optical systems are being devised, and a whole range of tactical countermeasures developed. While we cannot know which side will be ahead in 1980, 1985 or 1990, we can see that the current measure–countermeasure contest is a new game.

Before turning to the specific case of NATO, let us consider the effects these new weapons may have on the relative position of the smaller countries. Some years ago it would have been out of the question for most small countries to install a radar network and maintain *Nike*-size missile batteries to turn back enemy air attacks; in addition, batteries in exposed locations would not have had much chance of stopping a thrust by modern armoured units. The new style of arming goes a long way towards making the small countries more defensible on both counts. For some this will mean a new set of relations between the client nation and the larger power; in other cases the small power may have substantial wealth and much independence of action, and this may have an effect on the market for munitions. Perhaps many of these countries will find it in their interest to buy more anti-tank and anti-aircraft weapons, and fewer weapons more suitable for offence. With good fortune, the net effect in many regions may be a trend towards postures that are stabilizing.

Near-term effects of precision anti-tank weapons on NATO

How might NATO use these modern weapons in the 1970s against the 15,000 or so Warsaw Pact tanks opposite its Central Front?

On the Soviet side there will be the fierce anti-aircraft defences already mentioned and quantities of at least three anti-tank missiles, *Swatter, Snapper* and *Sagger*, or their descendants. The Soviet Union does not seem to have an air-launched anti-tank missile.[17]

---

[15] James R. Schlesinger, *Annual Defense Department Report, FY 1975*, Department of Defense, 4 March 1974, pp. 107–8, 152.

[16] For a more complete treatment of the implications of modern military technology see Kenneth Hunt, *The Alliance and Europe: Part II: Defence with Fewer Men*, Adelphi Paper No. 98 (London: IISS, Summer 1973), pp. 14 ff.

[17] None is mentioned in 'World Missile Yearbook', *Flight International*, 8 May 1975. However, General George S. Brown has said of Soviet air-to-ground operations: 'Airborne electronic countermeasure capabilities are being upgraded. More weapons can be carried, and new weapons, including tactical air-to-surface missiles, are being developed ... Five pylons have been observed in the *Flogger* C [a new variant with tandem seats]'. Of air-delivered precision weapons he said that, 'although our information is incomplete, we believe that we currently enjoy a considerable edge over the Soviet Union in the

On the NATO side there is a profusion of types of surface-launched missiles – with at least 16 due to be operational in the late 1970s. Nearly all these are wire-guided, and they include the previously mentioned *TOW*, *HOT* and *Swingfire*, with a maximum range of 3–4km, and the shorter-range *Dragon* and *Milan*. (For more detail, see Appendix I, particularly Table 2). Air-launched munitions include the laser-guided *Rockeye* (a cluster anti-tank munition) and *Maverick*, as well as helicopter-launched versions of *TOW*, *HOT* and *Swingfire* (for more detail, see Appendix I, Table 1).

I have already mentioned that the cost-per-round for *TOW*-class missiles is around $3,000, while the current procurement of *Maverick* has a unit cost of just under $10,000 per missile. Moreover, for automatic systems, like *TOW*, crews may be trained quickly, and there is no great problem of selection. Since, in addition, most of the systems mentioned are light and small, the number of PGM is likely to be legion. They can easily be adapted to be helicopter-mobile (though surface-launched) and should be natural candidates to serve as reinforcements or a *masse de manoeuvre*.

If all these potent properties of the new weapons are realized, it follows that there will be some new priorities on the battlefield. One of the biggest problems for all the systems mentioned above is target acquisition – though, once acquired, a target has a high probability of being destroyed unless it moves out of sight. I should therefore expect a war of seeing and hiding at the newly significant ranges of 2, 3 or 4km. If being seen at 3km leads to a high probability of being destroyed, there should be an increased use of smoke, camouflage and shielded paths for movements; equipment for night operations, and skill in using it, will also be important. This will be a competitive matter, in which the advantage goes to whichever side acquires targets at longer range.

Second, I expect both NATO and Warsaw Pact tactics will place a high priority on destroying PGM and air-defence units – by attacking either the crews or the equipment. This might be

attempted by barrages of anti-personnel artillery fire, by air-dropped weapons, or by trying to take launchers under direct fire.

These considerations lead naturally to a listing of some deficiencies of the present generation of anti-tank PGM, both Soviet and NATO.

(1) Although they need to be usable at night, in bad weather and in smoke and dust, almost all current PGM depend on the visible and adjacent parts of the spectrum for guidance.

(2) Launchers and crews are relatively vulnerable to artillery barrages and scatter bomblets.

(3) The rate of fire of most PGM launchers is lower than the probable rate at which targets would appear facing them in a typical Central Front situation. Also, the time of flight of most PGM permits evasion when targets see them coming.

(4) Many PGM use small shaped charges: damage from these can be repaired, or armour redesigned to withstand them. (*Maverick* is an exception. An Israeli colonel is reported to have complained about its performance in the October 1973 war, 'The damn thing blows up those Russian tanks so much that we can't fix them up for our own use.')

The defender can do something about most of these problems. For example:

(1) Long-wave infrared systems will work well on clear nights and fairly well in dust and smoke. Scout helicopters and electronic battlefield surveillance systems will help with target acquisition.

(2) Simple means of crew protection, such as operating anti-tank missiles from under armour, should not be expensive.

(3) For the time being a defender needs a mix of guns with high-kinetic energy rounds and missiles with shaped charges. The guns have a high rate of fire close in, where seeing is less of a problem, and armour redesigned to handle shaped charges may be vulnerable to high-kinetic-energy rounds.

In sum, from the NATO point of view the PGM of the next five years – the Class of 1980 – have their potential and have their problems, but many of the latter seem to be soluble at a tolerable cost.

Supposing that most of the necessary modifications are made and that anti-tank PGM are working at somewhere near their full potential, let us

military application of this technology' (*U.S. Military Posture for FY 1976*, Statement by General George S. Brown, Chairman of the Joint Chiefs of Staff, before the Senate Armed Services Committee, Washington, February 1975, pp. 104–6).

consider a simple numerical example.[18] Let us take a NATO division facing an offensive thrust by a Warsaw Pact tank army which has 1,000 tanks and many other vehicles. Our division's task is to stop 800 of the 1,000 enemy tanks, and it has at its disposal during the first two days of conflict:

    250 anti-tank land-based PGM launchers (*TOW, HOT, Dragon*, etc.)

    50 PGM-equipped helicopters, capable of flying 200 sorties

    50 fighters which can fly a total of 150 anti-vehicle sorties with 6 *Mavericks* per sortie

    Tanks, artillery, and mines sufficient to account for 100 enemy tanks.

The non-PGM weapons having stopped 100 tanks, let us assume that the total of 350 sorties by helicopters and fighters takes out another 400- (plus numerous other vehicles). Our 250 land-based PGM launchers must then stop 1·2 tanks per launcher to deal with the remaining 300 tanks (and, because the launchers would themselves be under intensive attack, these kills must be made while the defending units are still effective.

It is not the intention of this Paper to engage in detailed speculation as to the practicality of killing 1·2 tanks per land-based launcher, or as to the kill rates per helicopter or fighter sortie. However, if such a tank-killing capability *were* possible, then a very different and more hopeful picture of NATO's defensive potential will emerge in comparison with past estimates.

IV. PGM AND NUCLEAR OR CHEMICAL WARFARE

A major requirement – perhaps the major requirement – of American non-nuclear forces is that they preserve national interests without undue risk of escalation to nuclear war. For a long time there has been very little prospect of the United States or Britain embarking upon a *pre-planned* course of events leading to the use of even limited numbers of tactical nuclear weapons. There continue to be indications that the Soviet Union and France are similarly unlikely to engage in acts which involve high risks of nuclear conflict.[19] Thus, among the great dangers of military confrontation one must include situations which deteriorate rapidly, which are unexpected, and which could lead to dire consequences. This chain of events could be the more dangerous if communications were unclear, bluffs were misunderstood, or leaders were inept. Secretary Schlesinger has pointed specifically to the dangers of a general nuclear war having its origins in a deteriorating situation

<hr>

[18] The following example is very much a creature of its assumptions, and so is *not* a forecast.

[19] See the chapter by Thomas W. Wolfe in Kurt London, ed., *The Soviet Impact on World Politics: A Symposium* (New York: Hawthorn, 1974). Also Hannes Adomeit, *Soviet Risk-Taking and Crisis Behaviour: From Confrontation to Coexistence?* Adelphi Paper No. 101 (London: IISS, Autumn 1973). On the attitudes of French leaders see Marc Ullmann, 'Security Aspects in French Foreign Policy', *Survival*, November/December 1973. This discussion stresses the French emphasis on the political values of their nuclear forces. But French officials talk freely of early nuclear use in a deteriorating situation.

in Europe; the dangers would be multiplied if that situation had involved actual use of tactical nuclear weapons.

With these matters in mind, what can we say about the effect of precision-guided weapons on the probability of nuclear war? This is a subject which deserves full and separate treatment, but it is only possible here to raise a few points.

(1) The fact that defensive PGM are very potent (or could be made so) could go a long way towards stopping a tank thrust without resorting to tactical nuclear weapons. The potency of PGM and RPV in attacking targets in rear areas will be quite high by the 1980s and may substitute for medium-range nuclear strikes.

(2) On the other hand, a single nuclear warhead in the kiloton range might damage 100 vehicles dispersed over several square kilometres; to cause equal damage with present PGM might require 100 successful non-nuclear weapons. Non-nuclear area weapons are possible, but their technology is not yet well-developed, and they might be quite heavy. However, in addition to providing area coverage, tactical nuclear weapons have also been considered to compensate for inaccurate guidance and poor target location, and for this PGM do provide an adequate substitute. Some types, like RPV, can even home in on targets which have moved.

A more complete consideration of this point

would compare the effectiveness of precise non-nuclear weapons and nuclear weapons, with effects tailored for each of the several categories of target: dispersed and soft, dispersed and moderately hard (like tanks), hard point targets, targets of imprecisely known location, etc. From a purely technical point of view, some targets could be most effectively attacked with nuclear weapons and some with non-nuclear weapons.

(3) Some Western writers have considered the use of precision guidance to permit effective use of sub-kiloton nuclear warheads – so-called 'mini-nukes'. In part, these weapons were advocated because it was believed that the American strategic nuclear threat has become so decoupled from events in Europe that any American tactical nuclear contribution should be designed to repel an attack at its outset, and that early release of small yields would be more credible and result in less collateral damage. However, critics have questioned the physical effectiveness and nuclear efficiency of sub-kiloton weapons and have been concerned about the blurring of the nuclear firebreak.[20] If both sub-kiloton weapons and highly effective non-nuclear weapons were available the question for NATO would not be: Which can do the military job? It would be: Which (by possession or use) gives the signal most consistent with NATO goals?

(4) A strategy which aims at terminating or de-escalating a conflict will profit from the execution of precise, predictable, and understandable combat operations. The criteria for damage must be that damage to intended targets should be maximized, and to non-targets minimized. Non-nuclear PGM meet these requirements and in a great many cases serve a conflict-limiting strategy well. On the other hand, the consequences of even the most limited use of nuclear weapons are quite unknowable, and this makes their use risky

if the goal is to limit conflict.[21]

(5) We discuss below how modern weapons are likely to increase the rate of destruction in non-nuclear war, as well as the rate and cost of munitions consumption. This faster rate could lead to surprises, or to a pause where there might be some temptation to escalate to nuclear use. However, if we have anticipated the pause, and especially if we have been observing limitations in the preceding fighting, the pause could lead to de-escalation and negotiation.

Some Soviet strategic writings seem to regard non-nuclear war as a phase preceding nuclear war. If non-nuclear PGM do, in fact, halt a Warsaw Pact tank thrust, how will the Soviet Union calculate the value of going nuclear at that time? Two points seem straightforward. First, operating in small, separated units is a good tactic for PGM warfare and for avoiding tactical nuclear vulnerabilities. Second, passive protection and dispersal of rear-area support facilities is a good idea in any case. And to diminish both kinds of vulnerability is to decrease the other side's incentive to use nuclear weapons.

But the most important point is that the technology for precision delivery has come at just the time when Western strategy is turning towards the threat of carefully controlled combat – both in regional and in intercontinental conflict – as a more credible deterrent than the threat of unrestrained response.

Another topic which can only be mentioned here is the way the introduction of chemical warfare might affect these conclusions about PGM defences. There have been a number of warning signals in Soviet writings that they take chemical war preparations seriously. The Soviet-supplied chemical warfare equipment carried by captured Arab troops during the October 1973 war led to a public statement by the US Defense Department on the need for defensive preparations. At the least, this should make Western force planners look favourably on PGM systems that can be operated from enclosed vehicles or bunkers.[22]

---

[20] On 23 May 1974, the United States made a statement to the Geneva Disarmament Conference which 'gave assurance . . . that it would not develop a new generation of miniaturized weapons on the battlefield'. Dr Fred C. Ikle, director of the US Arms Control and Disarmament Agency, said in an interview: 'We have no intention to move in a direction that could blur the distinction between nuclear and conventional arms' (*New York Times*, 24 May 1974).

[21] See Albert Wohlstetter, 'Threats and Promises of Peace: Europe and America in the New Era', *Orbis*, Winter 1974, pp. 1122 ff.

[22] For a more complete discussion, see General George S. Brown's statement in *U.S. Military Posture for FY 1976, op. cit.* in note 17, pp. 114–120.

V. CONCLUSION

Subject to the several qualifications and omissions which have been pointed out, three main conclusions emerge.

First, the advent of PGM is probably advantageous to the defender. Target acquisition is the key to their successful use, and it is much easier for a defender to hide than for his opponent, who is moving through unfamiliar terrain and has no opportunity to prepare positions. Moreover, PGM, being relatively light, can be moved quickly to where they are needed – perhaps by helicopter – while heavier systems, including tanks, might arrive too late. Current PGM are not well suited to attack, since most of them are designed for specific defensive tasks; hopefully, therefore, the acquisition of PGM by both sides will lead to a more stable situation.

However, it is necessary to be cautious about concluding definitively that these developments always favour the defender (some years from now new target-acquisition systems, longer-range PGM and area munitions may help an attacker). Moreover – taking the NATO case again – even a massive Warsaw Pact offensive may involve the attacker holding defensively along 95 per cent of the front, while thrusting offensively along the other 5 per cent. And NATO forces must, in many places, go locally on the offensive. Weapon trends which assist efficient holding, therefore, will also suit Warsaw Pact purposes in the majority of places. The problem is to overcome NATO's capability to hold defensively in the remaining few places – where great concentrations of Warsaw Pact offensive strength would be needed. Again we come back to the question that requires detailed study: just how vulnerable to PGM are such concentrations?

A second conclusion is that an important consequence of the dispersal of so much destructive power down to small units, and the natural delegation of authority to use it, will be that the pace of war will be faster. In places with large concentrations of forces there will be an unprecedented intensity of non-nuclear conflict. Even though, as noted earlier, the total weight of munitions to do a job may decrease over the entire time of the conflict, the rate of use – in terms of proportion of stocks consumed – is likely to increase. The material destroyed is likely to be something like ten times greater

than we have been thinking about for non-nuclear war. We had a glimpse of this in the sudden logistic demands of the October 1973 war, but a war in Europe could dwarf those consumption rates. Will this pace lead to escalation or negotiation, when forces find munitions and equipment largely spent after three or four days?

Thirdly, there is a hopeful sign that the trend of the first part of this century, towards the inclusion of non-military target systems and civilian populations in military campaigns, will be reversed. Precision delivery means that military targets can be destroyed with less total explosive power and less collateral damage to non-military targets. The faster pace discussed above means that tactical forces-in-being, as well as strategic forces, count for more, and the general economy for less, in achieving a favourable outcome.

It is interesting to hypothesize that, given precision weapons, it may now be possible, with little loss of military efficiency, to adhere to a rule which strictly limits civil damage in both offensive and defensive operations. I should like to see this hypothesis explored in a broad analysis, and – if the result is favourable – it could be an appropriate subject for international discussion and negotiations leading to an agreed limitation.

Finally, another prospect for an arms-limitation agreement should be considered: one which seeks stability through an emphasis on defensive capability. Expensive, large, multi-purpose weapons (tanks, fighter bombers such as the F-14, nuclear aircraft carriers) are usually well-suited to offence. To the extent that smaller PGM-equipped units are making such systems less viable, it could perhaps be demonstrated that both sides would be served by limiting the numbers of such large systems. For a given budget or manpower ceiling, more resources could go into defensive units that would hold so well against an attack that, a generation from now, service school graduates may quote a new maxim: 'The best defence is a good defence'.

I must end on a note of caution. More precise long-distance delivery and the packing of more destructive potential into small packages is resulting in the rearrangement of traditional categories of weapons and systems (Proposition 7, page 82. At the same time there is increased interest in limited and carefully controlled operations, even at intercontinental range. For example, in the American forces, the long-range bombers of Strategic Air Command may be adapted to launch precision non-nuclear weapons while the aircraft of other commands may be armed with new weapons which make them more interchangeable with the 'strategic' bombers, except in range and payload. Arms-control agreements relating to 'strategic forces' will need to be replaced by agreements more directly related to the physical capabilities of delivery vehicles and their payloads, supplemented by agreements on the restrained use of forces.

Agreements on vehicles or substantial physical systems cannot be vitiated in the short-term, nor be responsive to short-term stimuli. But those which call for restraint in operations must stand up to the harder test of still seeming mutually desirable under the many short-term stimuli in the haze of war.

Appendix I

Current PGM and Guidance Techniques

Tables 1, 2 and 3 list some of the more significant PGM for use against land targets and aircraft; many of the weapon types which have caused the current wave of interest in PGM are included, but not all the types listed have been in large-scale production.*

A glance at Tables 1, 2 and 3 shows that a large number of systems in widespread use depend on the transmission of light in the visible spectrum (or, in the case of those which use lasers, just outside it). Thus many systems do not work well at night or in adverse weather (clouds, fog or haze), and their range could be limited by battlefield smoke or the deliberate use of obscurants. On the other hand, when conditions permit, visible-spectrum systems provide good resolution (and high angular accuracy), can use many inexpensive and widely available aids (such as telescopes, prisms, and television equipment), and present pictures in a familiar form for human operators to recognize and understand. The closer such a picture is to real-life scenes, the easier a man can use all his well-developed powers of visual observation. In fact, almost all current target acquisition for PGM involves a man looking at a scene or a pictorial representation of a scene, recognizing a chosen target among other things, placing a cross-hair or cursor over it, and pulling a trigger or actuating an automatic seeker. However, countering visual recognition is a familiar military task, and may not be very expensive. Camouflage, smoke screens, visual decoys, etc., can all be effective.

---

* For a brief treatment of almost all current guided missiles, see 'World Missile Yearbook' in *Flight International*, 8 May 1975; for a more extensive treatment, with many diagrams and photographs, see R. T. Pretty and D. H. R. Archer (eds), *Jane's Weapon Systems, 1974–1975* (London: Jane's Yearbooks 1974). For the technically trained, *Aviation Week* gives frequent detailed reports on progress in development and production of PGM systems, with emphasis on engineering choices, production problems and funding.

After launch the PGM may be guided by a homing device (for example, one which directs a missile towards the hottest spot in its field of view) or by commands from the launch point (the simplest systems fly the munition as if it were a model aircraft). A third type of guidance directs the munition (normally a missile) to a set of grid co-ordinates, which may be map co-ordinates or temporary electronic co-ordinates. This last type, which is usually not sufficiently accurate to qualify as 'precision' guidance, is not noted in Tables 1–3 but will be discussed in the section on mid-course guidance on page 100.

Particularly where fixed targets are involved, many countermeasures can be overcome by using area correlators (also called map-matchers) which compare a reference picture taken during previous reconnaissance, perhaps from considerable altitude, with the current picture seen from the missile. Since it will be impractical for a defender to change all the shapes recorded on the reference picture, the usual countermeasures will not work. This kind of guidance can give high accuracy, even though the missile may have travelled a long way and its launch position may not be precisely known. On the other hand, the prior reconnaissance mission does have to survive, afford recognition and produce a usable reference map. Correlators can be used with infrared or microwave sensors also, and for the latter their anti-jamming properties are even more necessary. A simpler system of the same sort can be used when the target itself is camouflaged or hard to recognize: the missile can be guided with respect to an offset aim-point. This requires that the attacker know both his own and the target's distance and azimuth from some easy-to-identify terrain feature.

However, the main impetus to use the non-visual spectrum comes from the need to operate in darkness and in bad weather. The options available will be discussed in Appendix II.

Table 1: Selected Air-to-Surface PGM for the late 1970s

Designation	Developed by	Range (km)	Weight of round (lb)[a]	Guidance	Comments
AS.20	France	7	315	Radio command	
AS.30	France	12	1,150	Radio command, some automatic features	Can be used against hard points and ships
AS.37 *Martel*	France/ Britain	30–60	1,170	Passive radio frequency seeker	For attacks on radars
AJ.168 *Martel*	Britain/ France	30–60	1,215	TV link + radio command	Can be steered by landmarks till target comes into TV camera range
Maverick AGM-65A	USA	b	460	TV tracker	Tracks automatically after lock-on
Laser-guided *Rockeye* KMU-420	USA	c	500	Laser homing	Primarily for anti-tank
Modular-guided Glide Bomb	USA	80(?)	2,000	TV or other + radio command	Evolving programme related to MK84-*HOBOS*
Shrike AGM-45A	USA	12–16	390	Passive radio frequency seeker	Used against ground radars
Standard ARM AGM-78A/B/C/D	USA	725	1,795	Passive radio frequency seeker	Used against ground radars
MK82-LGB KMU-388/B	USA	c	500	Laser homing	Steerable bomb
MK84-LGB KMU-351/B	USA	c	2,000	Laser homing	Steerable bomb
MK84-*HOBOS* KMU-353A/B	USA	c	2,000	TV homing	Steerable bomb
MK84-IR *Paveway* KMU-359/B	USA	c	2,000	Infrared homing	Steerable bomb
Bulldog AGM-83A	USA	10	598	Laser homing	Derived from older *Bullpup*
Condor AGM-53A	USA	110	2,110	Electro-optical/ TV homing or command	Remotely piloted
Walleye I	USA	c	1,100	Electro-optical/ TV homing	Carried by F-4, A-7, etc.
Walleye II	USA	c	2,330	Electro-optical/ TV homing or command	Used against large, semi-hard targets: e.g., bridges, ships

[a] In this table only, to agree with common American and British usage, weights are given in pounds. Bomb weights are nominal and actual weights may be 10 per cent more or less.
[b] Aerodynamic range has been given as 22km, but practical range for guided launch has not been released.
[c] Free fall.
SOURCES: *Flight International*, 14 March 1974; *Jane's Weapon Systems, 1973–1974; Aviation Week*, 10 December 1973, pp. 13 ff. and 15 July 1974, pp. 265 ff.

Table 2: Selected Surface-to-surface PGM for late-1970s Anti-armour Use

Designation	Developed by	Min./max. range (m)	Weight of round (kg)	Guidance	Comments
ENTAC	France	400/2,000	12	Manual command	Production complete, 13,000 produced
SS-11/AS-11	France	350/3,000	29·9	Manual command	Anti-submarine version for helicopters, 160,000 produced
SS-12/AS-12	France	800/6,000	76	Semi-automatic	Anti-submarine version for helicopters 8,000m max. range
HOT	France/ Germany	75/4,000	22	Semi-automatic command	May be used in Bö-105 light helicopter
Cobra	Germany	400/2,000	10·3	Manual command	No connection with the American AH-1Q *Cobra* helicopter
Milan	France/ Germany	25/2,000	6·7	Semi-automatic command	Two-man crew
Swingfire	Britain	< 150/4,000	34	Manual command + aids	Operator can be offset 100m, vehicle in defilade
Shillelagh MGM-51C	USA	(?)	27	Semi-automatic command, infrared link	Fired from 152mm gun mounted on the M-551 *Sheridan* or M-60A2 medium tank
TOW BGM-71A	USA	65/3,750	19	Semi-automatic command	Many carriers, but M-113 APC and AH-1Q *Cobra* predominate in US Army. (See note under *Cobra* above.)
Dragon M-47	USA	?/1,000	6·35	Semi-automatic command	Man-portable
Snapper AT-1	USSR	500/2,300	22	Manual command	Used in several Pact armies. Mounts on BRDM armoured reconnaissance vehicle.
Swatter AT-2	USSR	(?)	about 20	Manual command, infrared terminal guidance	Mounts on APC and BRDM reconnaissance vehicle.
Sagger AT-3	USSR	500/3,000	11	Manual command	Mounts on APC and BRDM reconnaissance vehicle (which carries 6 under a retractable plate).

SOURCES: *Flight International*, 14 March 1974; *Jane's Weapon Systems, 1973–1974*.

Table 3: Selected Surface-to-air PGM for the late 1970s

Designation	Developed by	Altitude limits (m)	Guidance	Comments
Crotale	France	50/3,000	Multi-mode	
Roland	France/ Germany	15/3,000	Optical (Roland I) radar or optical (Roland II)	Mounts on Marder or AMX-30; adopted by US Army in January 1975
Rapier	Britain	5,000 (?)	Optical or with Blindfire radar	Mounts on two Land Rovers
Blowpipe	Britain	(?)	Optical + remote control	Shoulder-fired
Redeye MIM-43A	USA	(?)	Optical aiming infrared homing	Shoulder-fired
Chaparral MIM-72A	USA	(?)	Optical then infrared	Modified Sidewinder 1C, typically used with Vulcan gun
Stinger XFIM-92A	USA	(?)	Optical aiming, infrared homing	Shoulder-fired, in development to replace Redeye
SA-6 Gainful	USSR	about 18,000	Radar or optical	Mounted in threes on a tracked transporter
SA-7 Grail	USSR	50/1,500	Optical aiming, infrared homing	Shoulder-fired or in batteries on trucks

SOURCE: *Flight International*, 14 March 1974.

A variety of factors influence the choice of wavelength to be used in a guidance system. Here we will discuss what is given up (especially in accuracy) and what is gained (especially in seeing through obscurants, including murky weather) by guidance systems not using the visible spectrum (usually defined as the wavelengths from 0·4 to 0·7 microns*). Again, this is a discussion for the general reader rather than the technical specialist.

The parts of the electromagentic spectrum most often used for self-contained guidance systems are:

Infrared: around 0·9 microns, 3 to 5 microns, and 8 to 14 microns.

Millimetre wave: around 8·6mm (35 GHz), 3·2mm (94 GHz), and 2·1mm (140 GHz)

Other microwave: 3cm (10 GHz), 6cm (5 GHz), and 10cm (3 GHz).

A rough idea of the loss in accuracy as wavelength goes up can be gained by considering what happens to angular resolution (the smallest angle between two equal point targets at which the viewing device can tell there are two targets, not one). Accuracy generally varies with resolution, but also depends on a complex set of other factors – for example, the signal-to-noise ratio and how long a radar beam stays on the target. Angular resolution is generally given by

$$R = \frac{\lambda}{D} k$$

where R is the resolution in radians

λ is the wavelength in centimetres

D is the aperture or antenna diameter in centimetres

k is a constant depending on system design and operation

For a given D, therefore, the angular resolution increases (i.e., deteriorates) proportional to wavelength. Thus, for example, a 3cm radar with a 75cm

* One micron, increasingly called a 'micrometre', is one-millionth of a metre. In the radio-frequency bands frequencies in Gigahertz (GHz = 10⁹ cycles per second) are often used instead of wavelengths. The approximate relationship between the two is

$$\lambda f = 30$$

where λ is the wavelength in centimetres
f is the frequency in GHz.

antenna diameter can typically get angular resolutions of 2° to 3°, but a 30cm radar would need a 750cm antenna. By contrast, the eye can resolve angles to about a minute, or 0·017°. Practical results with good visual bombsights show that cross-hairs can be set to within about 0·06°.

Infrared systems will be treated in the most detail below, since they can be used in a wide variety of terminal-guidance devices and are likely to be in widespread use by 1980.†

First, however, let us consider how often non-visual systems would be needed, on the basis of some data on central European weather.

Surface visibility and ceiling in central Europe

Central European weather, particularly in the region from the Ruhr to Stuttgart, is characterized by ground fog, industrial haze and frequent low cloud ceilings. R. E. Huschke has analysed a large body of data, taken over a number of years at several places in central Germany, and some of his results are shown in Figures 1 and 2.

Using data from standard visibility measurements made by observers looking out to see if they can distinguish familiar large landmarks, Figure 1 shows how the range at which a moving tank can be detected varies with surface visibility. Assuming the tank is continuously visible, H. H. Bailey's target acquisition model,‡ the two curves show detection range both for the unaided eye and for an observer with 6-power binoculars. The bars at the bottom of the figure show how often one can expect various surface visibilities; for example, winter visibility would be at or under about 2·5km a quarter of the time, and under 5km half of the time. Note that for these two examples the practical detection ranges of just above and just below a kilometre are much less than the 3km maximum range of several anti-tank PGM.

The curves in Figure 2 show how often air-delivered munitions can be used under two different assumptions as to ceiling and visibility; they also show the extent of daylight hours. These curves, devised by

† The remainder of this appendix draws on material originally prepared by S. J. Dudzinsky, C. C. Chen, L. G. Mundie, R. E. Huschke and P. A. CoNine, all of The Rand Corporation.

‡ H. H. Bailey, *Target Detection through Visual Recognition: A Quantitative Model* (Santa Monica, Calif.: Rand Corporation, RM-6158/1-PR, 7 February 1970).

Figure 1: Relationship of detection range to visibility in central Germany in daylight hours

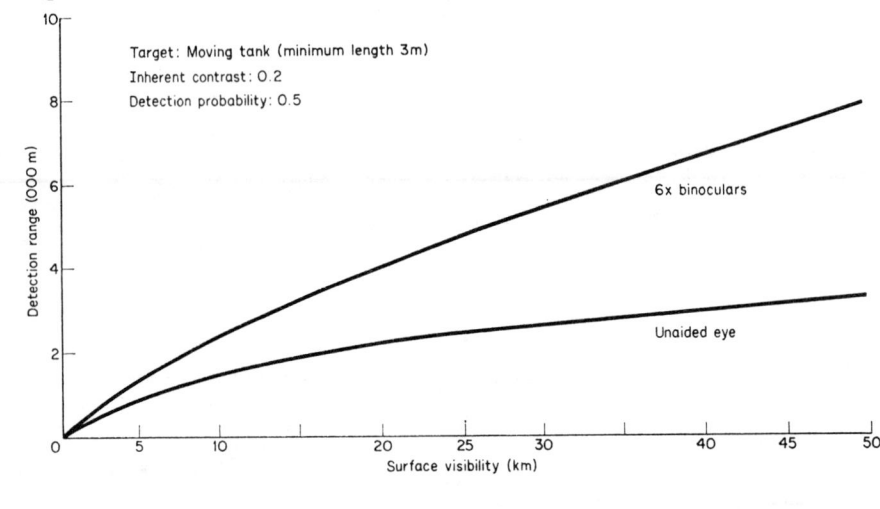

Target: Moving tank (minimum length 3m)
Inherent contrast: 0.2
Detection probability: 0.5

6x binoculars

Unaided eye

Detection range (000 m)

Surface visibility (km)

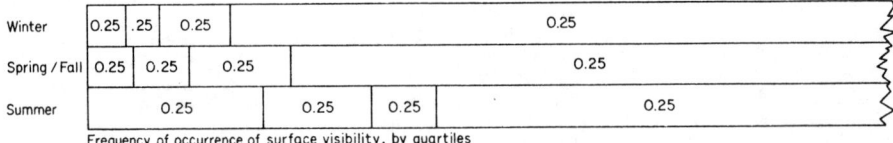

Winter	0.25	.25	0.25	0.25	
Spring/Fall	0.25	0.25	0.25	0.25	
Summer	0.25		0.25	0.25	0.25

Frequency of occurrence of surface visibility, by quartiles

Figure 2: Probability that ceiling and visibility values meet given limits near Berlin

(a) Ceiling 10,000 ft or more and visibility 4 miles or more   **(b)** Ceiling 1,000 ft or more and visibility 3 miles or more

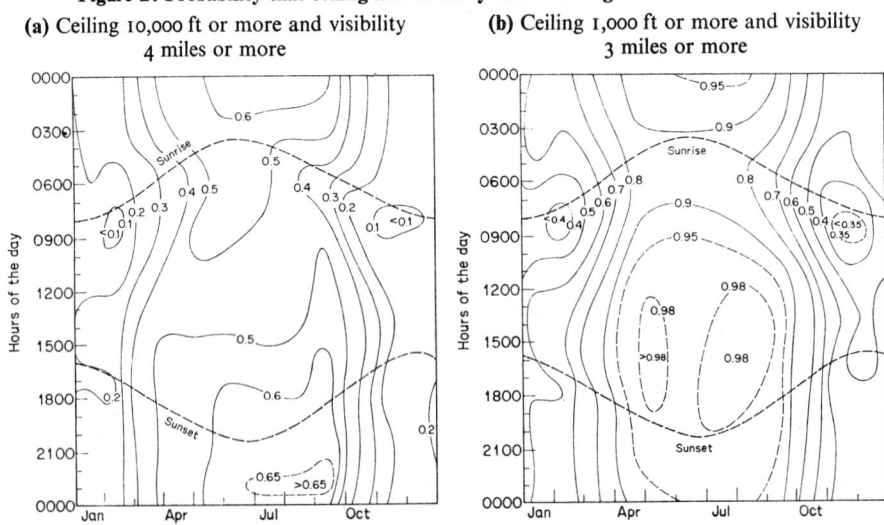

Figure 3: Transmittance over a 6,000-ft horizontal path at sea level in a clear atmosphere containing 10 g/m³ precipitable water

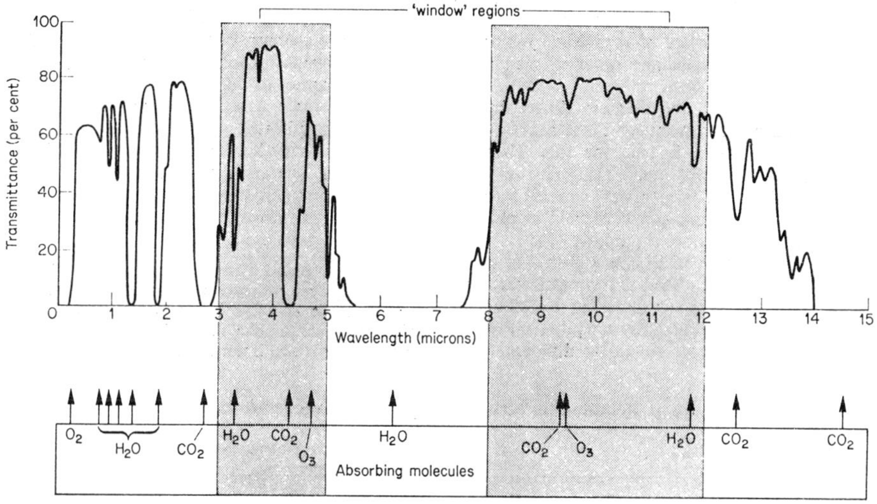

SOURCE: Adapted from R. D. Hudson, Jr., *Infrared System Engineering* (New York: John Wiley, 1969).

Huschke, are a bit difficult to follow at first glance, but they do provide a great deal of useful information. Figure 2(b), for example, shows that on a July afternoon the combination of 1,000-ft ceiling and 3-mile visibility can be expected between 95 and over 98 per cent of the time. At sunrise in December, however, it could be expected only about 40 per cent of the time, and there would be 16 hours of darkness.

Infrared systems for PGM

The infrared (IR) portion of the electromagnetic spectrum includes wavelengths from about 0·7 microns to more than 100 microns. Detectors designed for use above 3 microns often use the apparent temperature difference between target and background to get a useful signal. Thus, they can be passive, homing on the target's own emissions of radiation, and can detect targets under night-time as well as daytime conditions. Although such detectors could be developed for use in any portion of the IR spectrum, most effort has been devoted to development in the 3–5 micron and 8–14 micron 'window' regions, where attenuation of emitted energy due to atmospheric absorption is at a minimum.

The transmittance (proportion of energy transmitted) at the relevant wavelength over 6,000ft in clear atmosphere (i.e. no fog or haze particles) is shown in Figure 3. However, attenuation by water

vapour absorption can be significant even in the 'window' regions, and one can see from the diagram that, assuming the water vapour content of the atmosphere is 10 grammes per cubic metre, the average transmittance over the 8–12 micron region is about 70 per cent. (The water content mentioned corresponds to 60 per cent relative humidity at 20°C.) With a similar clear atmosphere containing less than 5 gm/m³ of water vapour (typical of mid-latitude winter conditions) and over 15 gm/m³ (representing warm, moist summer conditions) average transmittance would be about 85 per cent and about 55 per cent respectively.

One must add to the effect of water vapour the effect of scattering by particles in the atmosphere (e.g. by haze and smoke). If the diameter of particles is less than about one-tenth of the IR wavelength, then such attenuation is almost negligible (typical haze particles range from about 0·5 micron to under 1 micron, and phosphorous smoke particles are about the same size). However, fog and cloud (which are water particles) do pose problems for IR systems, because their sizes are comparable to IR wavelengths. (Attenuation by rain is not as severe, since raindrops are much sparser than cloud and fog particles.)

The effect of scattering on the detection range is shown in Figure 4 which, taking a particular type of cloud (stratus) and two types of haze, shows the relative value of 'generalized meteorological range' (i.e., detection range) as a function of wavelength. One can see that, in going from visible wavelengths to 10 microns, the gain in range (at constant humidity) is nearly two orders of magnitude for continental haze, and exceeds one order of magnitude for maritime haze.

On the other hand, the diagram shows that IR provides almost no advantage in seeing through clouds, and the same is true for fog. Thus, IR detectors, though useful under clear day or clear night conditions, and though they can see through considerable haze, smog and battlefield smoke, do not provide an all-weather capability.

Another implication of Figure 4 is that IR range is not uniquely determined by optical visibility. Over a period of time or series of locations, as obscuring particle size changed, IR equipment would work at greater or lesser ranges at different times

when the optical visibility was the same.*

To maximize detectivity, most IR detectors available today need cryogenic cooling: some to as low as 20°K (−253°C), although one of the most popular, the mercury-cadmium-telluride (HgCdTe) detector, only needs cooling to about 80°K (−193°C) the temperature of liquid nitrogen. Closed-cycle cryogenic coolers, which work for a long time, are generally complicated and expensive. However, if cooling is only needed for 10 minutes or so (the time required for operation of a detector on board a missile), a much cheaper blow-down cooler, which uses refrigerant stored under pressure in a refillable flask, can be used.

---

* Transmittance and scattering are not the only wavelength-dependent parameters to consider. For successful operation, the question is: Does the system have a sufficient signal-to-noise ratio? For a more complete treatment the reader is referred to Hudson, *op. cit.*

Figure 4: Relationship between range performance and wavelength

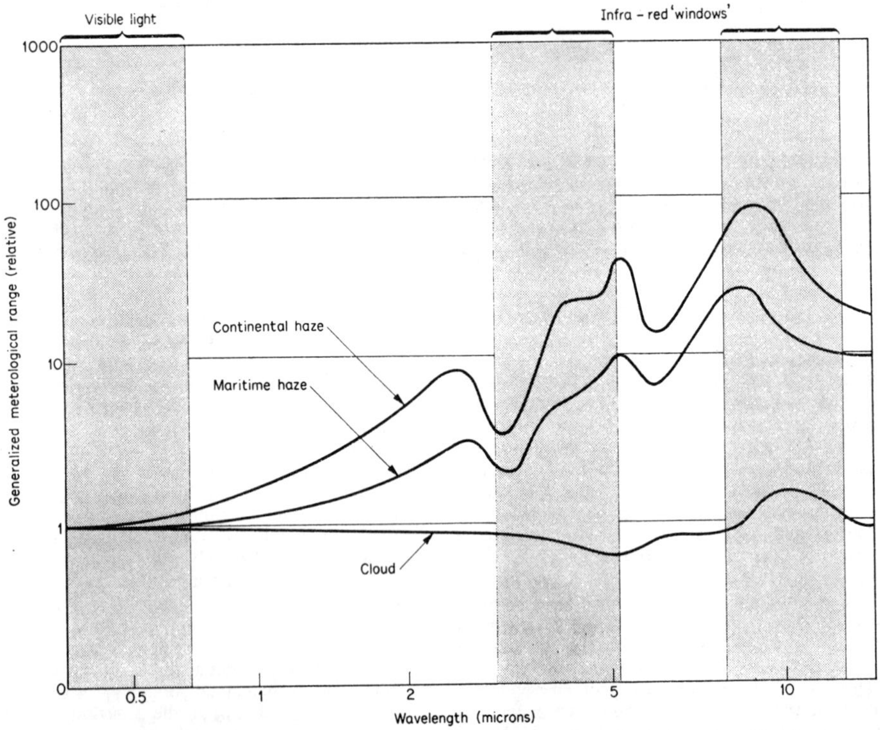

SOURCE: Adapted from D. Deirmendjian, *Electromagnetic Scattering on Spherical Polydispersions* (New York: American Elsevier, 1969) by R. E. Huschke.

Millimetre wave radiometry

One might expect electromagnetic sensors in the millimetre wave region to be useful in guiding PGM, since, by the equation on p. 18, modest antenna diameters would give excellent resolution by comparison with traditional microwave radars. One interesting kind of equipment to explore would be a radiometer (that is, a passive receiver) which would use directional antennas and superheterodyne receivers to detect thermal emission in the microwave region (though at these frequencies the observed signals obtained in field trials arise primarily from the emissivity differences between various objects, rather than from true temperature differences).

Microwave radiation penetrates clouds and fog reasonably well, and this is the main reason for seriously considering it in this application. Microwave radiometers are passive (hence their operation is covert, and their power requirement low), and they are relatively inexpensive and rugged. Their ability to generate good terrain pictures, even through fairly thick clouds, makes them useful for navigation and for attacks against fixed, pre-briefed targets, in conjunction with either manual control or area correlators. However, there are difficulties in using them for detecting and attacking moving targets.

The main weaknesses of practical passive microwave radiometers are the poor ratio of target signal to background clutter signal usually obtainable and the fact that they are relatively easy to jam. Figure 5 shows the probability of detecting a truck at a slant range of 3·6km for one set of reasonable parameter choices (including the assumption that the difference in emissivity between the truck and a field of grass which forms its background is about 0·9). At the assumed frequency of 35 GHz fog would have little effect over such a short transmission path; however, one can see that the antenna diameter needed for a 90 per cent detection probability exceeds 6 metres – much too unwieldy for most PGM applications.

The performance of radiometers operating at the two frequently used higher frequencies of 94 and 140 GHz is better than that of the 35-GHz radiometer in clear weather, since the antenna diameter required is inversely proportional to the frequency of operation. In certain adverse weather conditions, however, the increased atmospheric attenuation, the decreasing sensitivity of receiver circuits as frequency increases and other technical factors might combine to prevent satisfactory operation at these higher frequencies. Thus, millimetre-wave radiometers may not be satisfactory in detecting targets like trucks. They would still be useful, however, for navigation and for attacking fixed targets whose location in relation to major terrain features is known. Major terrain features would always be recognizable; the clutter problem does not apply, and obscuring them by jamming could be difficult.

Other microwave radar

Most guidance applications of microwave radar in the 'X' (3cm) or 'S' (10cm) bands are for anti-ship or anti-aircraft uses. Practical antenna sizes do not give the angular resolution needed to attack tanks

Figure 5: Probability of detecting truck with passive microwave radiometer at 35 GHz

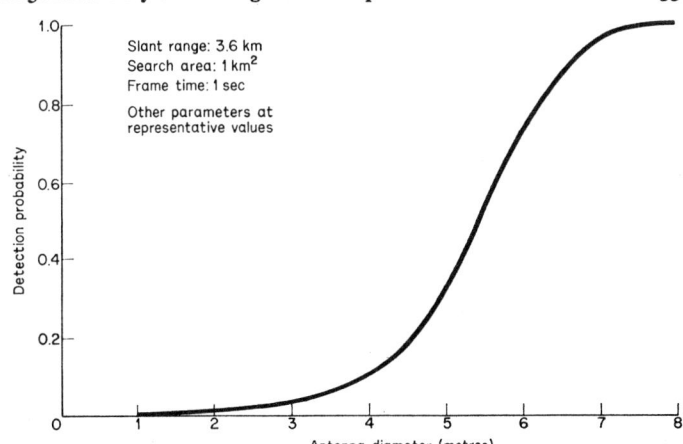

Slant range: 3.6 km
Search area: 1 km²
Frame time: 1 sec

Other parameters at
representative values

SOURCE: L. G. Mundie.

or trrucks. On the other hand, with ships and aircraft, not only is there great contrast between the target and its background but the target is usually surrounded by empty space. Anti-ship missiles tend, for a variety of reasons, to be large enough to take antennas of 30cm diameter or more; a simple radio altimeter can be used for guidance in the vertical plane; and the antenna beam need only scan around a single axis. For anti-aircraft use, the contrast of aeroplane against sky is even greater, while the missile is usually small, very manoeuvrable in response to guidance signals and is able to use a proximity fuse.

'X' and 'S' band radars are either active (with transmitter and receiver carried on the missile) or semi-active (the transmitter staying near the launch point to illuminate the target for an on-board receiver). Semi-active systems have advantages in keeping missile-weight low and in removing the transmitter's directly radiated 'spill-over' energy from the vicinity of the receiver. The latter is particularly useful where clutter from the ground must be filtered away in order to reveal echoes from the target, which is often done by using the doppler frequency shift. This shift, which depends on the velocity of the target, makes it possible to sort out and display only the echoes from fast-moving aircraft.

The accuracy with which most such systems can hit a target will depend on the pattern of radar returns reflected from the target, the aerodynamic characteristics of the missile, and the guidance accuracy of the radar. The dominant component of miss distance may often be due to the apparent shifting of the centroid of the echo as the relative position of target and radar changes.

Mid-course guidance for PGM

The three remaining systems we shall discuss have no strong correlation of wavelength with accuracy, so the wavelength can be chosen for good propagation over the required distance. While not accurate enough for terminal guidance in many cases, they could give mid-course guidance to direct a long-range PGM to a target area. It would then use terminal guidance (IR or TV, for example) for final precision guidance to the target. (The hypothetical 100-km cruise missile discussed on p. 3 would use mid-course, then terminal, guidance). The three mid-course guidance systems we shall discuss are:

Hyperbolic navigation of the Loran type, which is available today in many parts of the world.

A satellite navigational system (the earliest time for deployment of a fully operational complete American system of the type described is now estimated to be about 1984).

The combination of time-of-arrival and distance-measuring equipment systems, recently developed for attacks on missile guidance radars.

There are two basic modes of mid-course navigation: (1) an autonomous mode, in which the PGM receives navigation signals and processes them on board to determine its position and the required navigational corrections, and (2) a co-operative mode, in which the PGM retransmits the navigation signals it receives to a separate station where its position and navigational corrections are computed. Hyperbolic and satellite systems can use either.

In addition to the radio-frequency systems mentioned here, an inertial guidance device is sometimes used for mid-course guidance or for spatial reference. Inertial systems also often play a role in giving initial orientation to PGM, for example, as part of their airborne fire control systems.

Loran is one of several so-called 'hyperbolic' navigation systems. Precisely timed pulses are transmitted by two pairs of stations many kilometres apart. Lines of equal difference in the arrival time of the signal from one pair at the missile's receiver are hyperbolas, and points in space can be located by plotting the time differences from both pairs on a map over which the two families of hyperbolas are printed as a grid. Systems have been developed which automatically guide a missile to a point in the hyperbolic co-ordinate system, and this is accurate enough for mid-course guidance.

The US Department of Defense is currently developing a satellite navigation system: NAVSTAR Global Positioning System (GPS). The plan calls for three rings of eight satellites each in 12-hour circular inclined orbits, a configuration which ensures at least four well-located satellites would be in view from anywhere in the world. If the system works as planned, by 1984 highly accurate three-dimensional fixes will be available for aircraft, ships, ground vehicles and even troops. Positional accuracies are expected to be better than 30ft for 90 per cent of the time, and velocity can be measured to better than 0·2ft per second.

Both satellite and user equipment would have very accurate clocks, the user measuring the delays in the receipt of precisely timed signals, and thus the range to each of the four satellites and rate at which that range is changing. These data would be entered into a computer which would calculate the user's three-dimensional position. User velocity would then be determined by combining doppler measurements with the user-to-satellite position vectors.

An advantage of the system is that user equipment would be entirely passive, and thus would not give away its position to hostile forces; it is also said to contain a number of anti-jamming features.* The smallest of the user sets could be carried in backpacks by individual soldiers and is estimated to cost around $16,000, while 'Spartan' sets for a few thousand dollars, are a possibility. The user equipment for mid-course missile guidance would be similar but probably a bit more complex.

Time-of-arrival/Distance-Measuring Equipment (TOA/DME) was originally conceived by the US Air Force as a combination to detect and locate a radiating missile-guidance radar (using the TOA part) and then guide a munition to it (using the DME part). The TOA part of the system, designed to fix the source of radiation quickly (in case it is shut down to avoid further detection), locates active radio targets by noting the differences in the time of arrival of the same signal at two or more airborne stations whose separation could be measured by DME. The DME-guided weapon, which might be a glide bomb, is then launched. It carries a relatively inexpensive transponder (a transmitter which pulses when an incoming pulse is received) which responds to signals from each of two base stations. The round trip time of each signal and response from each base station to weapon and back is used to calculate range to weapon. This gives a circle of position from each base station, and the intersection of the two circles gives the location (the ambiguity caused by the fact that the circles intersect in two places can usually be resolved). DME can be used without TOA for guiding a weapon to a target located by other means. It is said to have an accuracy of about 100ft, enough to damage a soft target if the warhead is large. However, this kind of system might guide the weapon to the target area, when a terminal seeker (more accurate, but capable only of short-range use), perhaps using IR, would home on the target itself.†

---

* *Aviation Week*, 15 April 1974, pp. 22 ff.

† For further details, status reports, and discussions of combining various techniques see *Aviation Week*, 10 December 1973, pp. 13 ff. and 15 July 1974, pp. 265 ff.

4 Precision-guided Munitions and Conventional Deterrence

JOHN J. MEARSHEIMER

For ten years a revolution has been taking place in the realm of conventional weaponry, the principal result of which has been the proliferation of extremely accurate and therefore lethal weapons. The revolution centres on a type of weapon labelled precision guided munitions (PGM), although a number of key developments do not fall under the PGM rubic.[1] When PGM first began to attract public attention, some analysts claimed that these new weapons would favour the defence over the offence and thus enhance deterrence.[2] There was even talk about the 'death of the tank'. However, others pointed out that the claims made on behalf of these new weapons were greatly exaggerated and that they could be used effectively by both sides – and therefore they might weaken deterrence.[3]

The effect of PGM on conventional deterrence can best be understood by examining specific military strategies. On the modern battlefield, the essence of military strategy is how the offence and defence employ their armoured forces. There are two 'ideal' strategies between which an attacker can choose. First, the attacker can seek to defeat an opponent by engaging in numerous battles of annihilation, or set-piece battles. Ultimate success is predicated on wearing the defence down to the point where resistance is no longer possible. Second, the attacker can employ a strategy which is commonly referred to as the *blitzkrieg*. The mobility and speed inherent in an armoured force provide the means to defeat an opponent decisively without engaging in a series of bloody battles. The remarkable victories achieved by Germany in the early years of World War II, and decades later by Israel in the Middle East, amply demonstrate that total defeat of an opponent without resorting to numerous battles of annihilation is possible.

In a crisis, if one side thinks it can launch a successful *blitzkrieg*, it is unlikely that that side

will be deterred from striking. On the other hand, if the only strategy available to both sides is to engage in a series of set-piece battles, both sides will be very reluctant to attack. Certainly, one of the main reasons why Hitler had no reservations about striking against France was that the 1939 Polish campaign demonstrated that the *blitzkrieg* provided him with a superb weapon for quickly defeating an enemy.[4]

Therefore, the key question at hand is: what effect do PGM have on the *blitzkrieg* strategy? Do these weapons favour the attacker or the defender? Before considering these questions,

---

[1] Generally, a PGM is defined as a missile that is extremely accurate because it has a terminal guidance system. An example of a significant development in accuracy which is not PGM-related in a strict definitional sense is the greatly improved accuracy of main tank guns.

[2] See James Digby, 'Precision Guided Weapons', Adelphi Paper No. 118 (London: IISS, 1975); Colonel Edward B. Atkeson, 'Is the Soviet Army Obsolete?', *Army*, Vol. 24, No. 5 (May 1974), pp. 10–16; Colonel Stanley D. Fair, 'Precision Weaponry in the Defence of Europe', *NATO's Fifteen Nations*, Vol. 20, No. 4 (August/September 1975), pp. 17–26; and Kenneth Hunt, 'New Technology and the European Theater', in *The Other Arms Race*, edited by G. Kemp, R. Pfaltzgraff, Jr, and U. Ra'anan (Lexington, Mass.: Lexington Books, 1974), pp. 109–123. It is interesting that there is widespread agreement among Soviet defence analysts that these new weapons favour the defence. See Phillip A. Karber, 'The Tactical Revolution in Soviet Military Doctrine', unpublished paper, 2 March 1977.

[3] See James L. Foster, 'The Future of Conventional Arms Race', Rand, August 1975, P-5489; Richard Burt, 'New Weapons Technologies: Debate and Directions', Adelphi Paper No. 126 (London: IISS, 1976); Richard M. Ogorkiewicz, 'The Future of the Battle Tank', in Kemp *et al.*, pp. 43–55; Uri Ra'anan, 'The New Technologies and the Middle East: "Lessons" of the Yom Kippur War and Anticipated Developments', *ibid.*, pp. 79–90; and Jeffrey Record, 'Outwitting "Smart Weapons"', *The Washington Review of Strategic and International Studies*, Vol. 1, No. 2 (April 1978), pp. 83–85.

[4] See John Strawson, *Hitler as Military Commander* (London: Batsford, 1971), and Hans-Adolf Jacobsen, 'Dunkirk 1940', in *Decisive Battles of World War II: The German View*, edited by Hans-Adolf Jacobsen and Jürgen Rohwer (London: Andre Deutsch, 1965).

John Mearsheimer is a member of the Government Department, Cornell University.

two points are in order. First, even before the advent of PGM, the mere employment of a *blitzkrieg* strategy did not guarantee success. Contrary to the popular opinion of the time, the fall of France did not signify the ascendance of the offence. A defender skilled in the art of mobile armoured warfare can stymie an offence using a *blitzkrieg* strategy.[5] The issue is whether PGM *enhance* the defence or the offence, not whether they finally provide a means for thwarting the *blitzkrieg*. Second, this article will focus on those PGM which directly impact on the battlefield-weapons like *TOW*, *Sagger*, *Dragon*, *Milan*, *Maverick* and the various surface-to-air missiles (SAM). Long-range PGM like the cruise missile, that can strike important targets in an opponent's rear, and air-to-air precision guided munitions will not be examined.

PGM and the Blitzkrieg

The *blitzkrieg* depends on achieving numerical superiority at a point(s) in the opponent's front, piercing this front, and then following the path of least resistance into the enemy's rear. Although it may be necessary to engage in a set-piece battle to facilitate the initial breakthrough, a high premium is placed on avoiding further battles of this sort. The objective is to disrupt the victim's lines of communication and deny the defender time to reinforce weak points and regroup. The speed of the *blitzkrieg* denies the defender the capability to concentrate his forces for a second engagement. Follow on units can deal with the isolated strong points that the leading units of the offence bypass. As the battle develops, the initial armoured thrust can be augmented by flanking movements and pincer squeezes, although the central element of the *blitzkrieg's* success is the deep strategic penetration. However, this success does not rest solely on putting an opponent at a physical

[5] B. H. Liddell Hart, who, along with his British contemporary J. F. C. Fuller, was responsible for developing the theory behind the *blitzkrieg* in the 1920s, was arguing by the mid-1930s that a mobile defence could thwart a *blitzkrieg*. See B. H. Liddell Hart, *Europe in Arms* (London: Faber and Faber, 1937). In numerous publications after World War II, he provided extensive evidence to support his point that, even with the *blitzkrieg*, the relationship between the offence and the defence in World War II was not fundamentally different from that of World War I. For example, see chapters 10 and 11 in B. H. Liddell Hart, *Deterrent or Defence* (London: Stevens and Sons, 1960).

disadvantage. Psychological dislocation, which, of course, is a direct result of the defence's physical disadvantage, is the other pillar of success.

The revolution in precision-guidance has significantly enhanced the capability of a defence to thwart an offensive based on the *blitzkrieg*. Deterrence is enhanced because the number of weapon systems capable of destroying armoured vehicles has increased, and also because these systems are extremely accurate. In addition to using tanks and artillery, the defence can rely on shoulder-launched anti-tank guided missiles (ATGM); crew-served ATGM; helicopters and infantry-fighting vehicles equipped with ATGM; 'smart artillery'; and aircraft carrying a variety of 'smart bombs'. An attacking force that confronts a defender who has intelligently employed such weapons would have great difficulty making progress.

PGM present two problems for a *blitzkrieg* strategy. First, a mobile offensive requires the concentration of the enemy's armoured forces at a specific point(s) of attack to accomplish the initial breakthrough. Should the defender subsequently establish defensive lines, the attacking force would have to concentrate again to pierce them. Massing one's forces, however, is a dangerous tactic to use against an opponent armed with this growing arsenal of sophisticated anti-tank weapons. This is especially true in obstacle-ridden terrain where the attacking force is canalized. In this situation, only the offence's lead forces would be able to engage a defence armed with a plethora of lethal weapons. The remainder of the attacker's forces (the second and third echelons) would be unable to engage the defence directly – similar to naval forces that allow an opponent to 'cross the T'. Therefore, those few set-piece battles (especially the initial breakthrough) that the *blitzkrieg* has traditionally had to fight have become increasingly difficult to win. The great increase in fire-power resulting from the proliferation of new conventional weapons has raised the price that the offence must pay to pierce the defender's static front. Second, and more importantly, the notion of tanks pacing the attack, largely unsupported by infantry and artillery, is anachronistic in the face of a defence armed with PGM. The record of Israel's 190th Armoured Brigade in the 1973 Middle East War clearly demonstrates this point.[6] It is important to emphasize that historically the *blitzkrieg* has been propelled by armoured forces which did not have to concern themselves to any significant degree with supporting infantry and artillery. This does not mean that close co-ordination between the various combat arms was eschewed by practitioners of the *blitzkrieg*. Certainly, combined arms operations were necessary during the initial breakthrough and for subduing those defensive strongpoints which the main armoured force could not ignore. However, on the modern battlefield, the *blitzkrieg* will require armoured forces which are closely supported *at all times* by infantry and artillery. The tank is simply too vulnerable to operate unsupported, as it frequently did in the past.

Although the proliferation of systems capable of destroying tanks is the general cause of this development, the heart of the matter is the asymmetrical vulnerability of the tank and the individual soldier resulting from PGM. Before the revolution in conventional weaponry, the infantryman was a minor threat to the tank since anti-tank weapons like the 90 mm recoilless rifle were of limited value on the battlefield. Consequently, armoured columns driving deep into the defence's rear could virtually ignore pockets of infantrymen. Ensuing forces would deal with these threats. Now, however, tanks cannot ignore infantry strongpoints. Shoulder-launched ATGM like *Sagger* and *Dragon* as well as crew-served ATGM like *TOW* and *HOT* must be suppressed before the tank can advance. At the same time that the proliferation of extremely accurate weapons has done much to enhance the combat effectiveness of the infantryman, it has done little to increase his vulnerability. The PGM being deployed on the battlefield are designed primarily for use against weapon systems, not infantrymen. The same relationship obtains between the infantryman and an aircraft. An SA-7 or a *Stinger* represents a significant threat to an attacking aircraft, while a laser-guided *Maverick* missile is of little value against an infantryman.

The thrust of this argument should not be interpreted to mean that the battlefield of the future will feature a defensive force of PGM-armed infantrymen pitted against an offensive force dominated by tanks and aircraft. Certainly, any successful defence against a *blitzkrieg* will require large numbers of tanks and aircraft, as well as some type of infantry fighting vehicle armed with a PGM, a matter which will be discussed later. The key point is that a *blitzkrieg* places a high premium on armoured vehicles moving forward at a rapid pace, without having to rely extensively on infantrymen for support. A defence, on the other hand, relies heavily on both infantry and armoured vehicles. Given that the ability of the individual soldier to 'kill' armoured vehicles has increased significantly but the ability of the tank, or any PGM-armed vehicle for that matter, to kill infantry has not increased correspondingly, the defence benefits. In his autobiography, Moshe Dayan explains how this asymmetry manifested itself in the 1973 War:

> The principal combat factor was that in the north, most of the fighting took place with Syrian tanks on the attack and on the move, while our tanks were deployed in defensive positions. Thus . . . the Syrian *Sagger* anti-tank missiles had no special influence on the outcome of the battle.
>
> This was not the case in the south. In the first two days, our tanks were on the attack, hurrying toward the Canal, while the Egyptians – primarily infantry equipped with anti-tank missiles – were in defensive emplacements. And indeed, our tank losses in the south were caused by the defensive Egyptian deployment.[7]

This development can be contrasted with another advance in conventional weaponry. Over the past decade, the ability of tanks to kill tanks with their main guns has increased notably. Since both the offence and the defence rely heavily on tanks in *blitzkrieg* warfare, it is impossible to say which side benefits from such a development. That is not the case with the PGM-armed infantrymen; they clearly strengthen the hand of the defence.

The defence against a *blitzkrieg* is further strengthened because PGM permit a reduction in the size of the force necessary to hold a particular defensive line against an attack (the force-to-space ratio). The increased fire-power available to the individual soldier, coupled with such

---

[6] For a brief description, see Atkeson, *op. cit.*, p. 12.

[7] Moshe Dayan, *Story of My Life* (London: Sphere Books, 1976), p. 516.

developments as smart artillery, reduces the number of troops required to hold a front and releases them for use in a mobile reserve or for deployment as part of a defence-in-depth.[8] Other recent developments in conventional weapons technology which are not related to improved accuracy (improved conventional munitions, air-scatterable mines, fuel-air explosives, etc.), also contribute to the increased fire-power available to the defence and thus further lower the force-to-space ratio.

PGM clearly favour the defence when the offence is pursuing a *blitzkrieg*, thus making it increasingly difficult for an attacker to implement a *blitzkrieg* strategy. When an attack based on *blitzkrieg* principles fails, it evolves into a series of set-piece battles. As described earlier, this should work to enhance deterrence.

Rescuing the Blitzkrieg
A variety of arguments have been offered to rebut the claim that PGM have effectively eliminated the *blitzkrieg* as a viable military strategy. One frequently-mentioned panacea is increased co-ordination between armour and its sister branches – artillery and infantry. One analyst notes that 'infantry sweeps preceding armour may be a very effective means of dealing with a spread defence relying on PGMs'.[9] This strategy has little redeeming value. First, World War I demonstrated that the machine-gun makes infantry sweeps prohibitive. The vulnerability of the exposed infantryman has been further increased by the development of sophisticated anti-personnel devices. Second, once infantry is placed in front of armour, the notion of *blitzkrieg* warfare disappears. The mobility and speed of one's armour is then dependent on the pace established by foot soldiers. At this point, it is necessary to consider whether these anti-personnel devices which are so effective against attacking infantrymen can also be employed against the defender's PGM-armed infantrymen with equal effectiveness. The answer is clearly no, and the basis for this conclusion is that the attacker must move forward while the defence can fight from fixed positions. Therefore, the attacker's infantrymen will be either standing or

at best in a prone position for limited periods of time, while the defender's infantrymen will probably be in protected positions, or at worst in a prone position. An infantryman in a standing position is much more vulnerable than one in a foxhole.[10]

A more realistic solution would be to facilitate close co-ordination between simultaneously advancing infantry and armour supported by artillery and air power. The US Army refers to such a force as the 'combined arms team'. However, this approach also has important limitations. First, a co-ordinated attack involving such a diversity of forces is a complex task. The co-ordination of artillery fire with advancing infantry and armour is especially complex since mobile artillery does not have the luxury of making extensive firing preparations and to be effective its fire must be laid down as close to the advancing troops as possible. Second, whereas *blitzkrieg* warfare requires relatively little logistical support, a campaign based on the sustained use of a combined arms team approach would require a good deal more support. There would be a need for more ammunition, especially artillery rounds, and for more POL since the number of vehicles necessary to transport the infantry, the artillery and the ammunition would increase. Third, the maximum range at which PGM can engage targets varies from 1,000 metres for *Dragon* to 3,000 metres for *TOW*. Infantry, advancing simultaneously with tanks and armed with rifles and machine-guns, would not be able to engage PGM-armed soldiers effectively until the distance between them was somewhat less than 500 yards. Obviously, the PGM force would be at a decisive advantage since it would have first draw in the fight. Fourth, and most important, the pace of the attack would be slowed since tanks would still be consigned to keeping pace with advancing infantry. If heavy reliance was placed on artillery, the time spent preparing for and engaging in artillery exchanges would further hinder the rate of advance. On a battlefield ridden with sophisticated tank-killing systems, slowing down the speed, and therefore increasing the

---

[8] For an excellent discussion of how the force-to-space ratio has been decreasing, see Liddell Hart, *Deterrent or Defence, op. cit.*, chapter 10.

[9] Foster, *op. cit.* in note 3, p. 10.

[10] For example, *Army Field Manual 100–5* states 'Individual foxholes provide a 10-fold or greater reduction in casualties against impact fuzed artillery ammunition'. *Operations: FM 100–5* (Washington DC: Department of the Army, 1 July 1976), pp. 3–12.

exposure time of the tank, is clearly most undesirable.[11]

To combat the slowness and vulnerability of the dismounted infantryman, armoured personnel carriers (APC) have been developed and deployed. The objective is to develop a balanced attacking force using mechanized infantry that can keep pace with tanks and dismount only when necessary. These personnel carriers, which usually carry a squad of soldiers, are becoming increasingly sophisticated. They no longer are being designed merely to transport troops from point A to point B, but instead are being designed with enough fire-power, mobility and armour-plating to function as an 'infantry fighting vehicle' (IFV). This means that an infantry squad can conduct combat operations without dismounting. The Soviet Union began developing IFV and integrating them into her force structure well before the United States began actively pursuing the idea. The Soviet BMP, which is equipped with a 73 mm gun, an ATGM system and a coaxial 7·62 mm machine-gun, was originally designed to operate on a nuclear battlefield, where it was expected to exploit the many offensive opportunities resulting rom the use of nuclear weapons. Although the BMP is relatively thin-skinned (it was not intended for use against fixed positions), it is an integral part of Soviet strategy for a non-nuclear war.

However, evidence from the 1973 Middle East War indicates that IFV are very vulnerable on a battlefield dense with accurate anti-tank weapons.[12] This is basically the result of its armour, which is not as thick as the armour on a tank. An added disadvantage for the IFV is that a direct hit would probably result in the elimination of an entire infantry squad. Very importantly, the implications of the IFV's vulnerability

are different for the offence and the defence. Since an offensive force must move forward and since there are limits on the amount of protection afforded an attacker by the terrain, the attacking IFV will be very vulnerable to enemy fire. (This situation is the cause of Soviet concern.) As such, the IFV is of questionable value to the offence. It would make sense for an attacker to eschew the IFV concept and instead procure more tanks and build cheap APC that simply transport infantry from point A to point B.

However, because a defensive force usually fights from fixed positions, it is possible for an IFV to use man-made or natural obstacles for protection. Although this will not mean total invulnerability, an IFV's survivability is certainly greater in a defensive position than when it is rolling forward in the open as part of a strike force. This circumstance is to be welcomed because an IFV can significantly strengthen a defence. First, a defensive force on the modern battlefield will have to be mobile. Given the large size of PGM like *TOW*, it is necessary to mount a proportion of such weapons on mobile platforms. An IFV provides the capability to transport *TOW*s as well as infantrymen carrying shoulder-launched PGM. Second, an IFV affords infantrymen and mounted PGM a degree of protection when the attacker employs anti-personnel devices like artillery. The importance of this service cannot be exaggerated since an attacking force will undoubtedly use artillery, fuel-air explosives and other weapons to suppress infantrymen who threaten tanks. Clearly, the IFV favours the defence and not the attacking force. However, two caveats must be considered. First, a defence will be required to counter-attack – to go on the offensive – from time to time. In such instances, the value of the IFV to the defence becomes questionable. Second, in battles where the defence is not allowed to fight from fixed positions, battles where there is little distinction between offence and defence, the IFV is of little value to either side. In general, the IFV is a stabilizing system because it benefits the defence more than the offence.

The most serious threat to PGM is the development of 'special armour' or what is referred to in the West as 'Chobham armour'. Reports indicate that Chobham provides a three-fold increase in protection over conventional steel armour and that there is presently no PGM – Soviet or

---

[11] It is interesting to note that, as a result of PGM, for the defence, there is now less need for the infantry to rely on armoured forces for support, since infantry now have the capability to engage tanks; while for the offence, greater emphasis is being placed on integrated operations to deal with PGM-armed infantrymen.

[12] This development has been a cause of great concern for Soviet defence specialists. Phillip Karber writes that the Soviet Union 'had previously assessed APC to be twice as vulnerable as tanks. Apparently, in exercises and field tests since the Middle East War, the army has found that BMP is even more vulnerable to the new generation of anti-tank weapons than was previously believed'. Phillip A. Karber, 'The Soviet Anti-Tank Debate', *Survival*, May/June 1976, p. 108.

American – that can penetrate it.[13] Although there is no reason to doubt these reports, there are problems associated with this new technology. First, it is very expensive and, therefore, the number of armoured vehicles that can be outfitted with Chobham is limited. (Britain could not afford to incorporate Chobham into her *Chieftain* tanks.) For the foreseeable future, it seems highly unlikely that IFVs will be equipped with the new armour, and it would not be surprising if only a portion of a nation's tank force was protected by Chobham. Survivability has an expensive price tag. Second, although there are rumours that the new Soviet tank (T-80) might have special armour, all evidence indicates that the Soviet Union is behind the West in developing and deploying this technology. It will take a considerable amount of time for the Soviet Union to equip a significant portion of the Warsaw Pact's huge tank inventory with special armour. Third, while special armour is being developed and deployed, advances are being made in ATGM technology. Many of those systems used so effectively in Vietnam and in the recent Middle East War are essentially first-generation weapons. They represent the cutting edge of the PGM revolution. Future versions will be designed with Chobham armour in mind. Importantly, the speed at which technological innovations are incorporated into new generations of weapons favours the PGM over the tank. This is a result of the tank's greater complexity, which gives rise to technical problems and also tends to attract the kind of attention that inevitably results in a protracted development process. Although no one can predict developments in weaponry with great accuracy, there is no reason to believe that the effectiveness of PGM has been largely nullified by advances in armour protection.

However, should the balance continue to shift in favour of special armour, the battlefield equation would be significantly affected. The only PGM that would be capable of penetrating special armour would be the larger ones – like *Hellfire*. Shoulder-launched PGM and even crew-served PGM like *TOW* would be largely ineffective against vehicles equipped with special armour. Obviously, the value to the defence of such infantry-borne PGM will be inversely proportional to the number of special armour-equipped vehicles in the attacking force. If the trend is towards larger PGM, they will have to be mounted on IFV or some other mobile platform. As pointed out earlier, such a development certainly benefits the defence and not the offence. However, this does not negate the fact that the elimination of PGM-armed infantrymen as a key force on the battlefield would be detrimental to the defence. In general, it seems likely if advances in PGM technology are not forthcoming (other than to increase the size of the missile), the offence–defence equation will shift back towards the offence.

Given the rapidly escalating cost of increasingly vulnerable tanks, some argue that instead of procuring a limited number of expensive and sophisticated tanks, it would make more sense to deploy greater numbers of less expensive and less sophisticated tanks.[14] There are problems with this. First, inexpensive tanks are very vulnerable, and it is doubtful whether enough extra tanks could be procured to offset the higher losses that would result from the increased vulnerability of such an armoured force. The cost–exchange ratio between tanks and PGM clearly favours the latter. Second, the true cost of a tank force cannot be measured simply by multiplying the number of tanks by the hardware cost per tank. Tanks require manpower, and in North Atlantic Treaty Organization (NATO) countries at least, the cost of additional tank crews makes it very difficult to increase the size of one's armoured force. The British Army of the Rhine had to place approximately 50 *Chieftan* tanks in storage because they did not have the crews to operate them.[15] It is highly unlikely that the trend will be towards larger numbers of cheaper and less sophisticated tanks – for either NATO or the Warsaw Pact.[16] However, if the trend does go in that direction, it will certainly not threaten the utility of PGM like *TOW* and *Dragon*.

---

[13] Richard Ogorkiewicz, 'Tanks and Anti-Tank Weapons', in Adelphi Paper No. 144 (London: IISS, 1978), pp. 38–44.

[14] For example, see 'Critical Considerations in the Acquisition of a New Main Battle Tank', report prepared by the General Accounting Office, PSAD-76-113A, 22 July 1976.

[15] 'Rearming Without Tears', *The Economist*, Vol. 268, No. 7042 (19 August 1978), pp. 10–11.

[16] As a result of her war experience, Israel has concluded that it is premature to deploy well-protected tanks. See Merkava Mk 1', *International Defense Review*, Vol. 11, No. 7 (1978), pp. 1049–1052.

Responding to the IFV's vulnerability problem, some will undoubtedly argue for placing special armour on the IFV and effectively turning it into a tank with a missile instead of a gun. There is little utility in such a scheme since it is widely recognized that, for a variety of reasons, tanks should be equipped with guns, not missiles.[17] For the defence, the attraction of an IFV equipped with a PGM is its inexpensiveness relative to the cost of the tank. (These two systems also complement each other nicely.) Placing special armour on the IFV would effectively raise the price of an IFV to a level commensurate with the cost of a tank. The key assumption is that the protection the terrain affords the defender will compensate for the IFV's vulnerability. When the defence is forced to go on the offensive, primary reliance will be placed on the tank force. If there is a requirement for either the offence or the defence to increase its offensive punch, it would be more feasible to build additional tanks rather than place special armour on IFV. This is especially true when one considers the limited resources available to purchase armoured vehicles.

Some analysts argue that the attacker can negate the effectiveness of PGM by resorting to night attacks. There is abundant evidence that Soviet forces are well-trained in night operations.[18] However, there are problems with this approach. First, although it might be possible to achieve certain limited objectives with night attacks, it is hard to imagine the Soviet or any other military force inflicting total defeat on an opponent by relying exclusively on night attacks. The problems of co-ordination and poor visibility make such a strategy highly questionable. Furthermore, the assumption that PGM are ineffective in the dark because they cannot see the target is dubious. The United States is developing thermal-imaging night sights which will allow *Dragon*, *TOW* and *Maverick* to pinpoint targets in the dark. Finally, there is no reason why American forces, or any other defensive force for that matter, cannot be trained to fight at night. There is nothing inherent in night fighting which favours the offence.

PGM and the Air-to-Ground Balance

Both the German and Israeli *blitzkriege* relied heavily on close air support instead of land-based artillery for fire-power. Of course, the defence can use close air support to help thwart an armoured offensive. For both the defence and the offence the extent to which each side can rely on close air support depends on who has air superiority.

The deployment of air-to-ground PGM like the *Maverick* has greatly increased the combat effectiveness of close air support aircraft. At the same time, the effectiveness of ground-based air defence systems like SAM and air defence guns has also increased. In addition to evolutionary improvements in systems like *Hawk*, the SA-2 and the SA-3, new systems such as the highly mobile SA-6 and the shoulder-launched *Stinger* have appeared on the battlefield. The deployment of radar-controlled air defence guns like the ZSU/23/4 has further complicated the survivability problem for aircraft operating above the battlefield.

In providing punch for the offence, ground attack aircraft would be the ideal weapon, while the SAM and air defence guns would be used to protect the offensive force from the defender's ground attack aircraft. A *blitzkrieg* which confronts a defence that has neither an effective ground-based air defence system nor fighter aircraft could take maximum advantage of its ground attack aircraft. Conversely, ground attack aircraft that encounter a defence that has a belt of SAM and air defence guns, as well as a reliable fighter force, would have considerable difficulty assisting the advancing armour columns. The 1973 Middle East War demonstrated that a sophisticated air defence belt can exact a heavy price from attacking aircraft. Before the proliferation of mobile and accurate SAM and air defence guns, the defence had to rely on its fighter force and ground-based electronic countermeasures to counter the attacker's 'flying artillery'. Now it is possible for a defender, with no aircraft to speak of, to parry attack aircraft with SAM and air defence guns. Attacking aircraft confronting a defence that has both a potent fighter force and a sophisticated air defence belt would be largely ineffective. In such cases, the *blitzkrieg's* fire-power would have to be provided by land-based artillery.

An offensive force which has a well-integrated and mobile system of SAM and air defence guns

[17] Ogorkiewicz, *op. cit.* in note 13, p. 40.
[18] See Captain Eugene D. Bétit, 'Soviet Technological Preparations for Night Attack', *Military Review*, Vol. LV, No. 3 (March 1975), pp. 89–93.

would be well-suited to deal with a defence that relies heavily on ground attack aircraft. This is why there is so much concern among American policy-makers over the A-10, which is designed to deal with the cutting edge of the Soviet *blitzkrieg*. Of course, the problem that the offence encounters when it relies on ground-based systems to counter the defence's attack aircraft is that the size of the offensive force and its accompanying logistical tail increases.

What effect do these developments have on the offence–defence equation as it applies to the *blitzkrieg*? Assuming that the proliferation of the new conventional weapons technologies continues and that both the offence and the defence have ground-based air defence systems and close air support aircraft, the *blitzkrieg's* chances for success will be further complicated. First, when the defence deploys an extensive network of SAM and air defence guns, it becomes increasingly difficult for the offence to rely on close air support. This means that the main source of fire-power will have to be artillery.[19] This development is much more detrimental to the offence than the defence because it creates logistical problems which will work to slow the *blitzkrieg* and lead to set-piece battles. Second, the offence's reliance on SAM and air defence guns also creates significant logistical problems, which add to those logistical demands resulting from an increased reliance on artillery. Furthermore, the pace of the attacking armoured columns could be slowed since it is imperative that these forces do not outrun their air defence systems. This problem could be obviated by relying on fighter aircraft to provide air defence for the offensive forces and abandoning reliance on a ground-based air defence. For a variety of reasons, this does not appear to be the direction in which the major military establishments are moving. Third, in those instances where attack aircraft are able to bring their *Mavericks* and other sophisticated weapons to bear, the key problem will be target acquisition. This will be an especially acute problem on the European battlefield where visibility is limited during long periods of the year. The problem of target acquisition is more serious for the attacker's aircraft since the defence will be, for the most part, fighting from fixed positions. The offence, because it must abandon cover and move forward, will provide greater targets of opportunity for the defender's attack aircraft. Evidence indicates that the advances in weapons technology relating to the air-to-ground balance will contribute to the defender's capability to thwart the *blitzkrieg*.

Conclusion

As a result of the revolution in precision-guided technologies, it is much more difficult to implement a *blitzkrieg* strategy. To adjust to the proliferation of these weapons, the offence has been forced to increase the mass of his attacking force. An offensive must now place heavy reliance on artillery, SAM, air defence guns and mechanized infantry. The tank-dominated offensive, which relied on ground attack aircraft for fire-power support, has no place on the modern battlefield. The new emphasis on combined arms operations creates severe logistical problems as well as myriad problems of battlefield co-ordination, both of which rob the *blitzkrieg* of mobility and speed. The increased reliance that the offence is forced to place on artillery, to counter a defence depending on PGM, also contributes to the demise of *blitzkrieg* warfare. It is imperative to emphasize that the fundamental question is not whether PGM can be dealt with by an attacker, but instead, what changes in offensive strategy are necessary to overcome these weapons.

At the same time that PGM have compounded the attacker's problems, they have worked to benefit the defence. The increased fire-power available to the defence makes it possible to turn each major defensive position into a 'wall of fire' that the offence can penetrate only by paying an exceedingly high cost.[20] If a potential

---

[19] Since the 1973 war Israel, who up to now has relied heavily on close air support for offensive purposes, has significantly increased the artillery in her army. See Martin van Creveld, 'Two Years After: the Israel Defence Forces, 1973–75', *RUSI*, Vol. 121, No. 1 (March 1976), pp. 29–34.

[20] Regarding increased fire-power resulting from the proliferation of PGM, Mohamed Heikal writes concerning the 1973 Egyptian attack: 'It was this last-minute "overdose" of weapons (extra *Strellas* and *Saggers* given to the assault forces) that enabled the infantry to hold out, and General Dayan was later to admit that it was not so much the novelty of the weapons that took the Israelis by surprise as the sheer numbers in which they were available to the Egyptians at the outset of the battle'. Mohamed Heikal, *The Road to Ramadan* (New York: Ballantine Books, 1975), p. 6.

attacker perceives that using a *blitzkrieg* will evolve into a chain of set-piece battles, he will be very reluctant to initiate hostilities.

In practice, a large-scale offensive, be it a *blitzkrieg* or a strategy based on simply wearing the opposition down, is never purely offensive in nature. Any attack involves some combination of offence and defence. In those cases where the objective is to inflict total military defeat on an opponent, the attacking force is primarily concerned with the offensive ingredient. The *blitzkrieg* is such a strategy. However, when an attacker has limited aims, which would most likely involve the capture of some portion of an opponent's territory, defensive tactics assume much greater importance for the attacker. After a quick offensive surge, the attack moves to the defence and prepares for the opposition to counter-attack. The attacker uses the natural advantages that accrue to the defence, which are augmented by the proliferation of PGM. The victim, should he choose to launch a counter-attack, would be forced to attack a well-fortified and alert defence.

An offensive with limited objectives will undoubtedly attempt to utilize the element of surprise to achieve its objective before the defence has the opportunity to establish the 'wall of fire' described above. For the offence, surprise is a key means of dealing with PGM. Although surprise can provide the key to success in such limited operations, it should be emphasized that there are significant limits to the benefits one can expect to derive from surprise when the objective is total defeat of the opponent's armed forces. A defence that is versed in the fundamentals of mobile armoured warfare (unlike the Allies in 1940 and Egypt in 1967) will effectively halt any initial successes achieved by surprise.

The limited offensive stands in marked contrast to the *blitzkrieg* as a strategy which stands to benefit from PGM. The implications of this development are significant. Consider two examples.

In the Middle East, the value of an offensive based on limited objectives was clearly demonstrated by Egypt in the 1973 War. Israel suffered heavy losses in the first period of the war when she attacked solid Egyptian defences. It was only after Egypt abandoned this strategy on 14 October and launched an offensive that her position began to deteriorate. Even then, the margin between Israel's ultimate success and possible failure was precariously narrow. For Israel, who relies on a reserve army, the possibility of the Arabs achieving surprise and capturing some territory before the Israeli Defence Force can mobilize remains a real threat. Should an Arab state successfully pursue such a strategy, then Israel, who has traditionally relied on armour instead of infantry, must deal with a PGM-armed defence that is ideally suited to counter armour-heavy forces.

One would think that because NATO's defensive strategy is based on the concept of forward defence (i.e. thwarting a Warsaw Pact attack right at the border between East and West Germany), NATO's forces would be ideally positioned to deal with an offensive based on limited objectives. However, the majority of NATO's forces are located in peacetime well to the west of their defensive positions. Consequently, they will have to be alerted and then deployed to their forward positions in a time of crisis. In a crisis, however, NATO will be very reluctant to move its forces forward, since such a move could easily be interpreted as offensive in nature by the Warsaw Pact. This might trigger a Warsaw Pact attack that otherwise would not have taken place. On the other hand, if the Warsaw Pact believes that military action is inevitable, regardless of any NATO provocation, and they choose to pursue limited military objectives (an attractive alternative given the growing complications associated with the *blitzkrieg* and the Soviet aversion to engaging in a slugging match reminiscent of World War II), they have a vested interest in attacking before NATO can establish its 'wall of fire'. This means that in a crisis the Warsaw Pack will have an incentive to strike quickly – a destabilizing situation. NATO's best prospect for deterring such a limited strike is to deploy rapidly to the forward defensive positions. Unfortunately, it is very unlikely that NATO will know whether or not the Warsaw Pact is planning an offensive. Therefore, by moving towards the border, NATO runs the risk of provoking a attack – in those circumstances where the Warsaw Pact was not predisposed to launch an offensive.

In conclusion, although PGM have greatly enhanced the defence's capability to thwart a *blitzkrieg*, new problems have arisen regarding an offensive based on limited objectives.

5 PGM: No Panacea

DANIEL GOURÉ AND GORDON McCORMICK

In a recent article in this journal John J. Mearsheimer suggested that the advent of precision guided munitions (PGM) promises to 'revolutionize' conventional land combat. Mearsheimer argues that the enhanced accuracy of new, high technology weapons increases fire-power at the tactical level and that the character of these weapons, through placing a high premium on dispersal and concealment, principally compliments the assumed natural superiority of the defence. As a consequence, Mearsheimer concludes that the widespread employment of PGM will, for the first time, provide the defence with the means to counter the *blitzkrieg*, as 'classically' defined, thus enhancing conventional deterrence. Briefly stated, the greater fire-power of PGM when deployed in a defensive mode will blunt the high-speed armoured offensive by requiring the opponent to separate his massed forces, lest they be destroyed in the face of a now lethal defence. Dispersing offensive forces will slow the rate of advance, depriving the opponent of surprise and initiative, upon which successful *blitzkrieg* is premised. With offensive forces dispersed, combat will devolve into a series of 'set-piece-battles', which once again will favour defending forces by virtue of their prepared position.

The arguments are notable not because they are novel, but because they typify current thinking on the role and impact of PGM. While debate on the importance of PGM for the modern battlefield continues, the prevailing tendency is to view high technology munitions as a panacea for conventional vulnerability. Not surprisingly, attention has focused particularly on the potential role of PGM in West European defence, where fiscal and political constraints have promoted the quest for cheap and politically acceptable counters to a growing Soviet-Warsaw Pact armoured threat. While the threat is a real one, much of the discussion of defensive solutions through PGM is superficial and illusory.

Three issues raised by Mearsheimer and central to any discussion of the role of PGM in combat and deterrence deserve special attention: (1) the nature of *blitzkrieg*, (2) the inherent limitations of current PGM technologies, and (3) the adequacy of an attrition-oriented strategy.

What is Blitzkrieg?

A common fault of those who advocate the PGM solution, including Mearsheimer, is a misassessment of *blitzkrieg* doctrine and the armoured threat. In Mearsheimer's conception . . . 'the *blitzkrieg* has (historically) been propelled by armored forces which did not have to concern themselves to any significant degree with supporting infantry and artillery' (p. 70). He argues that after the initial breakthrough, armoured forces have been independently responsible for the remarkable success of this style of war. This notion, however, is incorrect. For it denies the very essence of *blitzkrieg* which has, from its conception, been the combined-arms approach.

Mearsheimer's confusion begins with a misreading of the classical origins of *blitzkrieg*. Specifically, the author fails to distinguish between the 'all tank' doctrine of J. F. C. Fuller and the combined-arms approach of B. H.

The authors are members of the Policy and Strategy Analysis Division of System Planning Corporation, Arlington, Virginia. This paper represents the views of its authors, not their employer.

Liddell Hart. Where the former became the leading influence on early British armoured doctrine, the latter's ideas were enthusiastically embraced by the original practitioner of *blitzkrieg*, the German *Wehrmacht*. While Fuller was correct in suggesting that tanks should be employed in spearheading an attack, rather than exclusively in reconnaissance or infantry support as in the prevailing French doctrine, he failed to recognize the need to diversify the armoured offensive with supporting arms. Liddell Hart made no such mistake. As early as 1924 he was calling for close co-operation between tanks, motorized infantry, mobile artillery, and aircraft for purposes of tactical breakthrough and rapid strategic penetration. It was Liddell Hart's firm conclusion that armour operating independently or with minimal infantry support could never achieve decisive results.

Blitzkrieg founded on the close co-ordination of tanks with supporting arms was the key German offensive operation throughout World War II. The individuals most closely connected with the formulation and initial implementation of German views on the subject were Guderian, Hoth, von Manstein and von Thoma. In each case support for the combined-arms approach was unequivocal. These views found operational expression in the Panzer Division, the organization of which has been well delineated by Richard Ogorkiewicz.

As well as having the tank brigade each panzer division also had from the very beginning a motorized infantry brigade whose role was to support and complement the tanks. The infantry, or rifle, brigade consisted of one 2-batallion truck-borne regiment and of one motor-cycle battalion. Each division also had a motorized artillery regiment with twenty-four 105 mm howitzers, an anti-tank battalion of towed 37 mm guns and an engineer company – quickly expanded into a batallion. There was also a reconnaissance battalion of armoured cars and motor-cyclists, as well as signal and divisional service units. The panzer division was thus a self-contained combined arms team in which tanks were backed by arms brought up, as far as possible, to the tanks' standards of mobility.[1]

While Mearsheimer is of course correct in stating that '... the mere employment of a *blitzkrieg* strategy did not guarantee success', and that '... the fall of France did not signify

the ascendency of the offensive,' the singular German victories during the initial stages of the war speak volumes on the value of fast-moving combined-arms operations as offensive method.

If the German army can be said to have originally codified the *blitzkrieg* as a viable offensive doctrine, then the Israeli and Soviet armed forces can certainly be said to have perfected it. In the case of the latter, the emphasis is once again on combined arms rather than the exclusive reliance on armour. Arguing against the fact of Israeli combined-arms doctrine, Mearsheimer cited the dismal record of Israel's 190th Armoured Brigade at the outset of the 1973 war. Yet here he makes rather selective use of the evidence, for it is difficult to picture a *blitzkrieg*, even in the author's definition, as consisting of merely one brigade. Combined arms have played an important and successful part in Israeli ground doctrine, as demonstrated by subsequent engagements in the Sinai, which were classic combinations of armour and supporting arms, including air power, artillery and motorized/ mechanized infantry. These efforts culminated in decisive victory on the ground. A particular incident, such as Sharon's 1967 victory at the Abu Agheila-Umm Katef cross-roads more correctly mirrors actual Israeli thinking and capability in the area of combined-arms operations.[2]

The Soviet style armoured offensive has taken the doctrine of *blitzkrieg* a step farther. While at the operational level, conventional ground doctrine emphasizes the familiar manoeuvre-dominated offensive, at the tactical level, the Soviet Union maintains her traditional predilection for mass attack, with the purpose of overwhelming enemy defences through sheer force of numbers and attendant fire-power. In this regard, though doctrine directs that offensive operations proceed in *blitzkrieg* fashion (i.e., the swift penetration of enemy rear areas with the object of defeating the opponent before he is able to mobilize), it equally equips Soviet ground forces to fight the set-piece battle should this become necessary in the course of the offensive.

Soviet offensive planning is based on her own version of the combined-arms army (*obshchevoy-skovaya armiya*), which reflects doctrine in providing the USSR with the means to engage in diversified ground combat.[3] Though always heavy in supporting arms, in recent years the

Soviet Union has significantly upgraded the ground forces, further improving her combined-arms capability. In the armoured division, manpower has increased from 9,000 to 11,000 men, while tank strength has risen from 280 main battle tanks to 325. Similar improvements have been affected in the motorized rifle division, where manpower has increased from 10,500 to 13,500 men and the number of tanks from 188 to 266, giving the division an armoured striking power equal to most NATO armoured divisions. Perhaps most notable in the light of emerging NATO PGM capabilities, however, are Soviet efforts to improve both armoured division and motorized rifle division fire suppression capabilities. In this regard, the numbers of towed artillery pieces have risen from 36 to 54 in the armoured division and from 105 to 144 in the motorized rifle division. These efforts are supported by improvements in the areas of multiple rocket launchers, surface-to-surface missiles and most important, by the development and deployment of self-propelled artillery. Finally, in addition to upgrading the traditional elements of combined arms, the Soviet ground forces have themselves been equipped with large numbers of PGM, though Soviet PGM capabilities generally fall short of those deployed by NATO.[4]

Thus, rather than, 'rescue *blitzkrieg* from the grave' as Mearsheimer rhetorically suggests it might, combined arms has, from the beginning, been an intrinsic feature of *blitzkrieg* operations. The traditional purpose of combined arms has been to provide ground forces with fire-power flexibility sufficient to overcome inevitably varied and unexpected battlefield problems. By insuring that the mobility of supporting arms is equal to that of central armoured forces, practitioners of *blitzkrieg* have successfully gained such flexibility without sacrificing manoeuvre capabilities. In addition to misunderstanding the character of *blitzkrieg*, Mearsheimer's optimistic assessments of the value of PGM to the defence are often rooted in a lack of appreciation for the limitations of current PGM technologies.

The Inherent Limitations of PGM

'Revolution in conventional warfare' is a phrase frequently found, and all too casually used, in the current debate regarding PGM. Real revolutions are few and far between; the introduction of gunpowder to the battlefield, the use of steam propulsion in naval warfare, and the invention of nuclear weapons and their union with the ballistic missile are perhaps the three genuine revolutions in military technology of the last one thousand years. Most so-called revolutions have less to do with new technologies than with their proper or novel exploitation. Although in 1940 the French were known to have tanks of superior quality to those of the *Wehrmacht*, it was the latter, based upon an innovative tactical and operational plan – the *blitzkrieg* – which brought about a 'revolution' in military strategy. Similarly, Israeli successes in both 1967 and 1973 against numerically superior Arab forces, armed with large numbers of both anti-tank and anti-aircraft PGM was due to the quality of the Israeli soldier, to the close co-ordination of armour, with infantry, artillery and airpower, and most importantly, to the superb strategic and tactical planning which took advantage of the elements of surprise and mobility.

It is a well-abused adage of the PGM 'lobby' that 'what can be seen, can be hit and what can be hit, can be destroyed'. Although the ability of these new weapons to destroy a target has certainly increased, the problem of target acquision remains. Most PGM require that the operator maintain 'line-of-sight' contact with the target during their entire flight time. Topographic and climatic conditions, however, often make such contact impossible.[5] In addition to natural impediments, the PGM operator may be subject to induced measures to limit target acquisition and tracking. Israeli tankers found several ways of degrading the effectiveness of Soviet ATGM, including the use of smoke, vehicle manoeuvering, and suppressive fire.[6] The latter was found to be particularly effective against PGM armed infantry whose munitions largely have ranges of 1,000 metres or less, putting them well within the killing range of tank-guns, much less artillery. The introduction of electronic counter measures (ECM) is another problem. US ECM equipment was highly effective during the latter days of the October 1973 war, in defending against Soviet SAM.[7] Newer model ECM equipment might be employed in countering anti-tank PGM.

Even assuming that the possibility of a hit remains high, structural and design changes to the tank, such as those included in the Israeli *Merkava*, promise to limit the killing capability of the PGM.[8] At the very least, such alterations might require the use of heavier, non-man portable PGM to re-establish lost capabilities. Such a development, however, would strike at the very heart of Mr Mearsheimer's argument; that the PGM favours the defence because of the ability to deploy it with infantry.

Another serious limitation of current PGM is their relatively slow rate of fire, a problem which becomes particularly severe in the face of a numerically superior and manoeuvre-oriented opponent.[9] Unlike artillery and tanks which are capable of firing every few seconds, the rate of fire of many PGM is only two or three rounds a minute. In this regard, in defending against a high speed armoured offensive, there will be many instances where targets simply appear faster than they can be destroyed.

The cavalier advocacy of PGM as the solution to the problem of conventional defence against the tank, fundamentally misinterprets the lessons to be drawn from their use, particularly in the Middle East. Indeed virtually the only case available to support the PGM 'lobby' is that of the unfortunate 190th Israeli Armoured Brigade which met disaster on 9 October 1973. A single engagement does not a revolution make.

The real instruction of the Middle East war of October 1973 is not to be drawn from the early slaughter in the Sinai of a single, unsupported, and incautious Israeli tank brigade by swarms of bazooka-wielding Arab infantry. ... The true lessons may be extracted from the knowledge that of the approximately 3,000 Arab and Israeli tanks destroyed or damaged during the war, at least 80 per cent were knocked out by other tanks. The use of hundreds if not thousands of anti-tank guided missiles (ATGMs) by both sides exerted at best a marginal influence on the outcome of the ground battle.[10]

The Strategy of Attrition

Colin Gray has correctly stated that NATO has two alternatives in preparing a conventional defence of Western Europe. NATO planners can either continue to implement a 'strategy' of forward defence based on halting a Soviet-Warsaw Pact armoured offensive at or near the point of attack, or they can abandon current practice and prepare for a counter-offensive war of manoeuvre.[11] Where the former 'strategy' seeks to engage the enemy in a war of attrition, the latter seeks victory not predominantly through

the destruction of enemy assets, but through the disruption of enemy operations and plans.

Most PGM enthusiasts, including Mearsheimer, fall into the camp of those who advocate a forward defence for NATO. The PGM has supported these advocates by appearing to promise unlimited fire-power with which to fight the battle of attrition on NATO's frontiers. Indeed, the argument for the massed deployment of PGM follows logically from a strategy of forward defence. Rather than justifying the inordinate reliance on PGM, however, in the light of their above mentioned limitations, this presents us with questions regarding the suitability of an attrition-oriented defence.

The notion of halting the Soviet *blitzkrieg* by means of attrition has proved a seductive one. However, it has tended to mask the otherwise ineluctable fact that a battle of attrition is necessarily brutal for *both* sides. Even in victory the costs to NATO could be staggering. Moreover, it is far from clear that NATO would emerge victorious, given the fire-power superiority of Soviet-Warsaw Pact forces which will *not* be eliminated by even the most extensive deployment of PGM. As Edward Luttwak has pointed out, the strategy of attrition must necessarily assume fire-power superiority, for 'in the cumulative destruction of the forces ranged against one another which characterizes an attrition contest, the inferior force will inevitably be exhausted first'.[12] Approaching the issue from the Soviet perspective, it is not evident why an aggressor, with numerical superiority sufficient to be confident of winning the set-piece battle of attrition, should necessarily be deterred from engaging in this type of offensive. Although the Soviet Union would not necessarily seek such an engagement, having initiated hostilities they would certainly accept the unappealing prospect of reducing NATO through attrition, should no alternative exist.

If good strategy is based on the identification and exploitation of enemy weaknesses, a static defence based on attriting the opponent (and oneself) is clearly not the preferred one. A strategy of attrition is one of uninspired generalship or desperation: this is the lesson of Grant in the Wilderness, Joffre at Verdun and Paulus at Stalingrad. Rather the military situation in Europe argues for a strategy which would take advantage of enemy vulnerability through

114

manoeuvre.[13] In this context, the role of PGM is further reduced. Mobility, intelligence and effective command and control will here be the key, not the PGM themselves.

PGM and the entire family of improved, conventional munitions serve to increase fire-power, and, with some note-worthy limitations, to establish a role for the individual infantryman on a battlefield still dominated by the tank and the airplane.[14] Yet, the PGM is no panacea. Firepower without a well-thought strategy is like a hurricane in mid-ocean: expending great energy yet wasting its force on the waves. More to the point, static defence based on *any* technology is inherently vulnerable. That is the now classic lesson taught by the French Maginot line. Historically speaking, PGM differ only to a slight degree from many of the anti-tank weapons used during World War II – notably the German 88 mm whose accuracy, range, rate of fire and shell weight made it, in its own time, an anti-tank weapon equal in effectiveness to many PGM. Yet, such weapons failed to achieve the extinction of the tank. It is strategy and the proper tactical application of both established and innovative technologies, which will prove decisive in future conventional engagements, as has been the case so often in the past.

PGM certainly have a role to play in modern war. However, what that role should be, and the extent to which they can be relied upon to buttress conventional deterrence, are issues which are still to be resolved. Mearsheimer's assessment of their battlefield impact is too optimistic.

NOTES

[1] Richard Ogorkiewicz, *Armour* (Stevens: Praeger, 1960), pp. 72–73.
[2] See Edward Luttwak and Dan Horowitz, *The Israeli Army* (Harper: Allen Lane, 1975), pp. 246–247.
[3] For a discussion of this aspect of Soviet operational/tactical doctrine see Steven Canby, 'The Alliance and Europe: Part IV, Military Doctrine and Technology', *Adelphi Paper*, No. 109 (London: IISS, 1975), p. 9.
[4] Further detail on recent Soviet Ground Force changes is provided in John Erickson, 'Trends in Soviet Combined-Arms Concept', *Strategic Review*, Vol. 5, No. 2 (Winter, 1977), pp. 64–79.
[5] Richard Burt, 'New Weapons Technologies: Debate and Directions', *Adelphi Paper*, No. 126 (London, IISS, 1976), p. 11.
[6] Charles Wakebridge, 'A Tank Myth or a Missile Mirage?', *Military Review*, Vol. 56, No. 9 (August, 1976), pp. 8–9.
[7] Mike W. Fossier, 'The Role of SAMs in Tactical Warfare', in *The Other Arms Race*, edited by G. Kemp, R. Pfaltzgraff, Jr. and U. Ra'anan (Lexington: Lexington Books, 1975), p. 38.
[8] 'The Merkava Mk 1.', *International Defense Review*, Vol. 11, No. 7 (July, 1978), pp. 1049–1052. Also Richard Ogorkiewicz, 'The Future of the Battle Tank', in Kemp, *et al.*, *op. cit.*, pp. 43–56. Possible changes to the battle tank include new composite armour, placement of engines in front compartments, increasing the oblique angle of the turret and hull armour, and use of non-reflective paints and insulation to foil laser and infra-red ATGM designators.
[9] James F. Digby, 'Precision Guided Munitions: Capabilities and Consequences', in Kemp *et al.*, *op. cit.*, p. 10.
[10] Jeffrey Record, 'Outwitting Smart Weapons', *Washington Quarterly*, Vol. 1, No. 2 (April, 1978), p. 83.
[11] Colin Gray, 'Nato Strategy and the Neutron Bomb', *Policy Review*, No. 7 (Winter, 1979), pp. 13–14.
[12] Edward Luttwak, 'The American Style of Warfare and the Military Balance', *Survival*, Vol. XXI, No. 2 (March/April, 1979), p. 58.
[13] For a discussion of the need for a manoeuvre strategy for NATO see Luttwak, *op. cit.*, and David K. Anderson, 'The Counter-Mobility Potential in the NATO Context', *Strategic Review*, Vol. 7, No. 1 (Winter, 1979), pp. 67–75.
[14] Mr Mearsheimer and others often ignore the fact that most PGM, particularly those of longer-ranges, are *not* man portable and require the use of vehicles such as the Soviet BMP.

6 Military Competition in Space

If either the United States or the Soviet Union were to be deprived of the use of military satellites in a time of crisis she would find herself at a marked disadvantage. Both have come to rely on the capabilities of satellite systems to provide the instantaneous warning and information needed in a contemporary war. Even in peacetime, the use of space for reconnaissance, communications, electronic intelligence, ocean surveillance and navigation has burgeoned. Surveillance satellites – the so-called 'national technical means' of verification – are essential to each superpower in order to ensure that it has detailed information on the other's strategic weapons, and they are thus the vital underpinning for agreement in arms control. Early-warning satellites are relied upon to identify missile launches, and electronic intelligence satellites can then track the missile to determine its ultimate impact area, thus assuring each side of the other's intentions. Communications satellites are vital for the near-instantaneous transmission of commands to units all over the world. Navigation satellites are already necessary for naval units (particularly

strategic submarine forces), and in the near future missile guidance will come to depend on satellites. Even fighting units on land and in the air will come to rely on satellite data for precise position definition.

Both powers have become deeply conscious that space-based systems are extremely vulnerable to counter-action. New technologies have raised the possibility of serious interference with, and even the destruction of, satellites, and this has forced both sides to examine ways of defending their space-based systems against a pre-emptive attack.

Events in 1978 drew attention to different aspects of the military use of space. On 24 January the Soviet ocean surveillance satellite *Kosmos* 954 crashed in a fortunately uninhabited area of the Canadian North-west Territory, spewing radioactive fragments of its nuclear reactor over a large area; and two days later the People's Republic of China launched its second reconnaissance satellite. During the year, public attention was drawn to the continuing research and development that both the United States and the Soviet Union were conducting into anti-satellite (ASAT) measures, ranging from exploding killer satellites close to their targets, via the launching of impacting fragments into a target satellite's orbit, to exotic techniques based on lasers and charged-particle beams.

As the launch of the Chinese reconnaissance satellite makes clear, it is not only the super-powers which are involved in military competition in space. The number of states which maintain their own space programmes, and of those which intend to use these programmes for military purposes, is growing and will continue to grow in the coming years. Although China is only the third power to launch her own military satellites, the time is not far off when this small club will be joined by others. India plans to make the first of her own launches, using her SL-B-3 rocket, in July 1979 (two Indian satellites have already been launched by the Soviet Union). One of the satellites to be launched is a remote sensing satellite, though whether its camera resolution will be sufficient for military application remains to be seen. NATO and British communications satellites have been launched by the United States, but no other NATO power has yet carried out a launch of its own. In 1983 France plans to launch a remote sensing satellite with some military reconnaissance capability (*SPOT*) and is preparing a dedicated military reconnaissance satellite which will be put into orbit in 1985. Japan has launched more non-military satellites than any other state except the USSR and the US, and certainly has the capability to build and launch military satellites if she ever felt it were necessary. The private German firm OTRAG is developing satellite launchers and testing them in a large rented area in Zaire; it intends to place its first small payloads into orbit within the next two years and hopes to achieve geosynchronous altitudes in the early 1980s. These new abilities to launch reconnaissance and communications satellites will soon compound the problem, since satellites which can be used to verify arms control can also direct a military attack.

Ocean Satellites

The crash of *Kosmos* 954 called attention to ocean surveillance satellites, an area of space technology in which the Soviet Union has followed a different path from that of the United States and in which she now has a significant lead. In December 1967 the Soviet Union began to launch satellites that exhibited very peculiar behaviour: after remaining in a low orbit of about 250–300 km for some days (and later for as much as two months), they split into two, one part burning up on re-entry into the atmosphere and the other being raised to a higher circular orbit of about 900–1,000 km. The reason was that the satellites contained an active sideways-looking surveillance radar which required a small nuclear reactor (containing about 50 kg of enriched uranium) for a power source. Once it ceased to produce enough power for the radar, the reactor was separated and boosted into high orbit, where it was supposed to stay for up to 900 years while the reactor materials decayed to safe levels.

The Soviet ocean satellite surveillance system appears to have become operational in May 1974, when for the first time satellites were used in pairs, with the second satellite following the first into the same orbit half an hour later. Not only could ships' positions be plotted but, by monitoring changes in the plot occurring within this time span, their speed and direction could be ascertained. Even the types of vessels could be deduced to a certain extent from their radar cross-sections. One such pair of satellites has been launched every year since 1974, with the exception of 1978 when, probably due to the *Kosmos* 954 accident, two pairs were launched. In 1974 the launches coincided with the NATO naval manoeuvre *Dawn Patrol*, and in 1975 with the Warsaw Pact manoeuvre *Okean 75*. If developed further, this system may provide real-time targeting information on US ships.

The United States is developing a comparable radar satellite system under the code name *Clipper Bow*, which is expected to be operational by 1983. Hitherto she has relied on maritime patrol aircraft and on passive electronic intelligence satellites under a programme known as *White Cloud*. These satellites are launched in clusters of four (two such clusters have been launched to date), and it is believed that one 'mother satellite' co-ordinates the information gathered by the three 'daughters' which monitor radar transmission from ships. Differences between the signal strengths received by the three 'daughter satellites' give the location of the ship, and the signal characteristics, or combinations of certain signals, will help to identify the kind of naval vessel. It is possible, for example, to identify *Nanuchka*-class missile corvettes from their radar emissions, because their radar system is not known to be used by any other Soviet naval vessel. However, since the system depends on detecting radio emissions, detection can easily be avoided by maintaining radio silence for the period of overflight.

Navigation and Guidance

The American Global Positioning System (GPS) is a system of twenty-four *NavStar* navigation satellites, to be placed in groups of eight at 20,000 km altitude, which is expected to be operational from the mid-1980s. Satellite navigation information is currently provided by four *Transit* satellites in polar orbits at about 1,000 km altitude. By monitoring two standard frequency signals and the orbital data they transmit, the position of a receiver can be determined during each overflight with an accuracy of some hundreds of metres. The main purpose of the *Transit* system is to update inertial navigation systems of nuclear ballistic missile submarines so as to ensure sufficient missile delivery accuracy, but other naval units also use the signals for navigation.

The limitations of the *Transit* system are that it only provides two-dimensional information, and this can only be updated on each satellite pass. Moreover, because the position-fixing depends on exact measurement of a frequency shift, the system cannot be used by fast-moving receivers like aircraft and is very vulnerable to electronic countermeasures (ECM). Also, it relies on only four satellites, and these are within range of the Soviet ASAT system.

All these problems will be solved by the GPS, which will enable military receivers to locate their position in three dimensions at any time and at any place in the world (including space, up to altitudes of a few thousand kilometres) with an accuracy of 10 metres. Velocity, it is claimed, can also be determined with an accuracy of 6 centimetres per second. The satellites' transmissions are protected against ECM and unauthorized use (by so-called spread-spectrum pseudo-random-noise signals), although another signal, providing a less accurate fix, will be made available to civil airlines and shipping. The GPS will improve the performance of all forces to a remarkable extent. For the first time there will be a common reference grid world-wide, which can be used for determining the exact relative position of all users. GPS receivers can be used not only by naval forces but also by ground forces or aircraft. Thus a target of which the co-ordinates are known can be attacked by aircraft flying blind, and reconnaissance aircraft will be able to correct navigation errors and drifts and follow predetermined flights paths with great accuracy. Moreover, there are no difficulties in principle to using GPS guidance even for ballistic missiles (including those launched from submarines), which would reduce their circular error probable to a few tens of metres over intercontinental distances. The GPS thus promises to be one of the most significant recent improvements in military technology.

Anti-Satellite Systems

The vulnerabilities of the US and Soviet satellite systems differ. On the whole, the Soviet Union deploys more satellites in low orbit, because she is unable, for geographic reasons, to launch the very high geostationary satellites used by the United States. In addition, the United States tends to use fewer satellites because she can use one vehicle to fulfil several functions. This makes US systems more vulnerable and has caused the United States to be very concerned about the continuing Soviet anti-satellite programme.

Tests of Soviet satellites designed to seek out and destroy other spacecraft were begun in 1967. The initial phase of the programme ended in 1971, but extensive testing was resumed in 1976. Most of the tests have been simulations. The one on 19 May 1978 was typical: *Kosmos* 1009, launched from Tyuratam, was initially placed in a transition orbit, but before completing a full revolution it was shot up towards its target satellite *Kosmos* 967, passing it at a distance of 1 km at an altitude of about 1,000 km. In some of the tests the interceptor has exploded into a large number of fragments in close proximity to the target, which is then destroyed by the impact. It is believed, in addition, that the Soviet Union is making preparations to launch a satellite-mounted high-energy laser which could be used to destroy other satellites.

The 19 May test illustrates both the capabilities and the limitations of the Soviet interceptor system. Although the orbits of interceptor satellites can reach altitudes of 2,000 km, all interceptions have taken place at altitudes of less than 1,000 km. Current American reconnaissance, electronic intelligence, meteorological, ocean surveillance and *Transit* navigation satellites are thus all threatened by the Soviet system. The time between the launch of the interceptor and the interception is very short (1–2 hours). Because the number of American satellites normally operational for these purposes at any one time is usually small (15) the number of interceptors needed is not great. Another group of satellites within the range of the Soviet interceptor system is the Satellite Data System (SDS), consisting of three spacecraft which provide communications over polar areas and are believed to act as relay stations for reconnaissance satellites flying over the Soviet Union. They are placed in highly eccentric orbits, which reach an altitude of nearly 40,000 km over the northern hemisphere, but come as low as 300 km at their perigee over the southern hemisphere, where they are vulnerable to interception. Also theoretically vulnerable is the American *Space Shuttle*. This can reach altitudes up to 1,200 km, but its usual operational altitude for placing spacecraft into orbit will rarely exceed 500 km.

There are, however, important limitations to the Soviet ASAT system. It is still impossible to reach the *NavStar* GPS satellites, which will orbit at 20,000 km, or the US communications and early-warning satellites, which are in geostationary orbit at 36,000 km. The interceptors cannot change their orbital plane and therefore can only be launched when the launch site lies in the orbital plane of the target satellite, and this only happens twice a day. But in only a fraction of these passes can the target satellite be reached by the interceptor within a few hours, since in the majority of cases when the orbital planes coincide the distance between target and interceptor is too great. Thus an interceptor launch is possible only once in several days. This has the important consequence that, whereas a single satellite might be intercepted by surprise, degrading a satellite system which uses several spacecraft might take a matter of weeks.

The United States has now begun to develop her own ASAT capability. The most promising project uses a small manoeuvrable warhead developed from antiballistic-missile technology. This is propelled by a relatively small rocket into the flight path of a satellite, where it releases a number of small missiles which home on the target using an infra-red terminal guidance system. The system differs from the Soviet one principally in that it is planned to be launched from an aircraft, which offers much greater flexibility of use. The launching aircraft can be flown into positions from which the targets can be intercepted at almost any predetermined time. A first test of the warhead in space is planned for 1981.

The range of this weapon is expected to be comparable to that of the Soviet system, which means that satellites can be intercepted at up to 1,000 km but that those in geosynchronous orbits cannot be reached. This limitation is less important for the United States, since the Soviet Union's more northerly position dictates that she cannot make use of geostationary orbits but must place her early-warning satellites and most of her civil and military communications satellites in highly eccentric orbits. Because of their high apogees, some 40,000 km above the northern hemisphere, they are visible from the Soviet Union for some 80% of the time, but the corollary is that during their perigees, low over the southern hemisphere, they will be within range of the American ASAT system. The Soviet Union normally maintains 16 *Molnya* communications satellites and three early-warning satellites in space at any one time, and all have to pass through the low perigee every twelve hours. If the American ASAT system becomes operational in the 1980s, it promises to be far superior to the present Soviet system, partly on account of its greater flexibility and partly on account of the greater vulnerability of the Soviet satellites.

Directed Energy Weapons
Both the United States and the Soviet Union are becoming increasingly interested in directed energy weapons, both high-energy laser (HEL) and particle-beam weapons (PBW). Some of the fundamental research on these is directed towards their use for controlled nuclear fusion or nuclear weapons simulation, so that the funding which appears in defence budgets is only a small part of the money actually being spent. Even so, the US defence budget for FY 1979 includes $184.1 million for laser weapon development and $11 million for particle-beam weapons (which makes it clear that US planners have better hopes for the former than for the latter).

High-Energy Lasers
Research into the military applications of high-energy lasers was initially undertaken by the US Defense Advanced Research Projects Agency (DARPA), but development has now been handed over to the services.

The properties of laser weapons (a straight beam moving at the speed of light, with short reaction times and high aiming accuracy) make them particularly suitable against aircraft, cruise or guided missiles or ballistic re-entry vehicles. The main problem with mobile, ground-based lasers (such as might be used for defence against incoming missiles, for example) is that, in anything other than clear weather, water and other particles in the atmosphere absorb so much of the beam's energy that the range at the moment is less than 10 km. This limitation can be avoided in the case of ground-based ASAT lasers; such weapons could be static, and this would allow use of larger, more powerful beams and they could also be set up in areas where clear weather is likely. This should enable them to achieve the 'lethal' range of about 700 km. Lethal ranges of space-based sys-

tems against satellites are difficult to estimate, and depend in part on the target satellite's vulnerability to laser radiation. Although a laser beam in the vacuum of space is not weakened by absorption, the beam spreads, and thus decreases in intensity, with distance. It is therefore difficult to envisage space-based ASAT laser systems with very long ranges. This difficulty is particularly important for a projected anti-ballistic-missile (ABM) system using high energy lasers which is being considered by the US Army. A space-based system would have the advantage that ICBM could be intercepted over Soviet territory in the boost phase, while they are large and vulnerable, with their propellant stages still attached. However, the limited range of lasers would not permit them to be placed in those high geostationary orbits from which a few could cover all the areas through which ICBM must pass; beam intensity would have decreased too much with distance to be certain of damaging the target. But if space-based ABM systems were placed in lower orbits from which they could inflict damage, e.g. at 1,000 km, they would only occasionally pass over areas where ICBM could be intercepted and would therefore spend most of their time in the wrong place. In order to overcome this, a large number of lasers would have to be maintained in orbit, which, apart from being very expensive, would create a highly exposed and vulnerable system.

As far as laser weapon systems are concerned, therefore, not only are there fundamental technical problems to be solved, but also it remains uncertain whether in the end they will be competitive with other weapon systems which are intended to fulfil the same function, such as missiles.

Particle-beam Weapons
This uncertainty is even greater with particle-beam weapons. It is known that the Soviet Union has been active in particle-beam research for many years, particularly with respect to controlled nuclear fusion and for industrial applications. Soviet particle-beam research is believed to be concentrated at Sarova, near Gorki, and at Semipalatinsk. Extensive activities (particularly at the latter facility) have led some

to conclude that a major Soviet breakthrough in beam weapons development has already been achieved. But this conclusion is doubtful.

Charged-particle beams consist of streams of sub-atomic particles (protons or electrons) accelerated to very high velocities in an electromagnetic field and then released in the direction of a target. Hydrogen ions can also, in theory at least, be used to form hydrogen atom beams. In both cases, the power required is enormous. Yet, compared to lasers, particle-beam weapons have the advantage of higher efficiency and less attenuation by atmospheric water content. They are therefore less dependent on weather conditions, and, even if the beam intensity is insufficient to damage a target structurally, the target's electronic components may be affected by the ionizing radiation produced by high-energy particles hitting its surface. Targets might therefore be more vulnerable to particle beams than to lasers.

But particle beams, too, have several inherent problems. Used within the atmosphere, the beam interacts with the air, causing a loss of energy and instabilities which make precise aiming difficult. The earth's magnetic field bends charged-particle beams and this adds to the deviation. In addition, the mutual repulsion of the beam's charged particles increases beamspread and the beam thus loses intensity with distance.

American particle-beam programmes are now being co-ordinated under the auspices of DARPA, in order to investigate whether some of the basic technical problems can be overcome. Thus far American efforts have been concentrated on three possible areas of application. Most advanced is the US Navy's *Chair Heritage* programme, which by the mid-1980s might lead to a first prototype electron-beam weapon of short-range (a few kilometres) for defence against cruise missiles. This might be based on large ships such as aircraft carriers. The US Army has been pursuing a concept aimed at developing a space-based electronically-neutral particle-beam weapon for ABM purposes. Such a system would also have ASAT capability. As with space-based laser systems,

however, a large number of these would have to be deployed. Moreover, because of their considerable size, each satellite carrying such a weapon would have to be assembled in space by means of several *Space Shuttle* flights. This would be a conspicuous and time-consuming operation which might allow the Soviet Union to act diplomatically, or even militarily, to prevent it. Such satellites would be very vulnerable to ASAT attack.

The third system, also under consideration for the US Army, is a ground-based proton-beam ABM system. This programme is still in a very early stage, and it is uncertain whether it will prove technically feasible.

Indeed, the technical problems associated with both laser and particle-beam research are such that it is not yet possible to forecast whether any of these ideas will in the end lead to weapons systems. If the problems (mainly, but not only, in the area of particle-beam weapons) can be solved, effective self-defence systems may prove feasible for ships and aircraft large enough to carry the heavy devices and the corresponding power facilities. In any event, laser weapon systems seem unlikely to be operational before the mid-1980s, and particle-beam weapons could not be ready until a good deal later than that – well into the 1990s.

Protective Measures

In parallel with the development of ASAT systems, countermeasures to the threat from the other side are being developed, tested and deployed on US spacecraft. To reduce a dependence for power on the very vulnerable solar paddles, two experimental communication satellites were launched in March 1976 (LES 8 and LES 9) which used radio-isotope thermo-electric generators. Efforts are being made to reduce electronic circuit sensitivity by shielding, by the choice of less sensitive components and by built-in redundance. So-called 'dark' satellites are being developed to reflect less light and reduce radar echoes, thus reducing the danger of detection. Impact sensors are being attached to US satellites to distinguish between interference by an aggressive act and an accident. Radar detectors have been built which warn against possible radar emissions from the terminal guidance systems of interceptor satellites. Horizon sensors used for altitude control can be protected by a shutter, and satellites under warning of attack can use additional propellant to manoeuvre away from the interceptor. A system for ejecting decoy satellites is also being developed.

This is an impressive list of safeguards, yet none of them ensures survivability of satellites. Rapid replacement may therefore be necessary. The US *Space Shuttle* could be used to place satellites in orbit, but in wartime conditions would itself be vulnerable. The USAF therefore plans a limited silo-launched capability for rapid satellite replacement by early 1980. To establish this capability, it is planning the trial launch of a satellite using a *Minuteman* III or *Titan* II missile.

Arms Control

Efforts to protect satellites include political as well as technical initiatives. On 8 June in Helsinki, the United States and the Soviet Union began talks on ways of restricting anti-satellite weapons. (The 1967 Outer Space Treaty only banned space-based nuclear and other weapons of mass destruction. It did not prohibit the use of a satellite or other space-based system to destroy another orbiting system.) The recent negotiations are an indication of the mutual recognition of the importance of space systems to the strategic balance; at the same time, the inability thus far to reach any form of settlement is a reflection of the political and technical problems confronting the negotiators. Yet the prerequisites exist for an agreement: neither country has a commanding lead in the relevant technology, which in any case is rudimentary and barely deployed, and technological breakthroughs are still a decade or more away. As a result, it is quite possible that both the United States and the Soviet Union might want to avoid the risk of falling far behind in this crucial technology more than they want to compete for the advantage a lead would bestow.

The primary obstacle to a comprehensive ban on anti-satellite capabilities is verification. Lasers can be easily concealed, and in some cases 'killer satellite' systems can be tested under the disguise of a normal satellite launch. Nor is there a clear line separating military activity from any other – the *Space Shuttle*, for example, could be used to place both military and civil satellites in orbit, and may have the capacity to capture or destroy enemy satellites. But there is a fundamental difference between the ability to take out one or two satellites without warning and the capability to destroy most or all of an adversary's satellites simultaneously. While the former would cause a serious political crisis, only the latter would threaten to give a clear advantage to one side. Neither super-power is as yet close to such a capacity. It is this which may induce both to seek an agreement before the other side has moved beyond the technological constraints which still apply today.

7 New Trends in Air Power

The nature of air power is changing. It can now be considered as the application of military force by or from a vehicle in the third dimension, up to and including outer space. This includes offence, defence, manned and unmanned systems, and transport and reconnaissance in the context of a national or allied defence policy. Air power can be supplied by servicemen in several different colours of uniform and, in its totality, involves manned aircraft, electronic warfare, command, control and communications systems (C^3), surveillance systems, remotely piloted vehicles (RPV) and terminally guided weapons.

New Technology
Today's and, especially, tomorrow's technology offers opportunities to air power: in improving the aircraft platform, enhancing target location, acquisition and destruction, in improving C^3 and, above all,

in applying electronic warfare on a larger scale than even before. Technology also increases the chance of survival in a hostile air environment.

Historically, offensive conventional fire-power has implied mass fire-power, largely because of its basic inaccuracy. In terms of air power, this means large formations of aircraft for even a small chance of success, and the weight of effort required to attack targets during World War II and the Korean War and in the early stages of the Vietnam War was formidable. Dependence on mass was the result of two related factors: first, little more than inspired guesswork was attached to the weapon release point, and, second, the weapon release parameters (height, speed and wind velocity) were not known with sufficient precision to 'aim' the weapon usefully. An example of how this affects force requirements can be seen from the planning figures for manually released bombs against a 60-metre steel-plate girder bridge. To give an 83% probability of dropping a span, no less than 160×450 kg bombs would be required.

Once the release point can be made precise, delivery accuracies can be improved significantly. Soon the introduction of systems like the GPS satellites, providing a forecast accuracy of 10 metres in all three axes, and of faster on-board inertial navigation and weapon-aiming systems will have a most profound impact on the application of air power. If terminal guidance were applied, weapon expenditure would be dramatically reduced. Using the example of the 60-metre bridge, only four laser-guided bombs would be required to ensure its destruction – an overall improvement factor of 40. Such bombs' delivery accuracy is itself also much less dependent on range from the release point.

The most significant impact of the new technology stems from developments in micro-processors and associated data-processing techniques. Not only have they revolutionized weapon accuracy, but their weight and size have been greatly reduced and no longer limit their usefulness. Traditional problems of discovering the enemy's position, disposition, strength and movement are being rapidly overcome by sensors using radar, radio, infra-red, optical and other equipment in manned and unmanned aircraft and satellites. In C³, advances in micro-electronics are permitting the transmission of more information of greater complexity over wider areas, at a faster rate and with cheaper equipment. Commanders at the very highest level can thus involve themselves in distant conflicts, while commanders at the lower level can be kept fully informed.

Electronic countermeasures (ECM) and defence against them – electronic counter-countermeasures (ECCM) – together make up the fastest-growing aspect of modern air warf. In contrast, the development of radar as tended to slow down, and although there have been many developments to enhance and improve its performance, it must be considered nearly at the peak of its development potential, while ECM technology is just getting under way. It is this latter capability more than any other which is likely to have a dynamic impact on the future of air power. Although an appropriate countermeasure will inevitably be found, any early advantage could prove decisive in a battle. Thus both East and West are attaching considerable effort and resources to the ECM/ECCM area.

There will be continuing development in aerodynamics, material technology and propulsion systems which will, for example, allow vertical or short take-off and landing characteristics to be coupled with satisfactory operational range. Another area in which rapid strides can be expected is that of the cruise missile. The great drawback to the early missiles was their comparatively short range and poor accuracy, which made them suitable for use only against large targets (such as London for the V-1 rocket) or as nuclear delivery vehicles where the size of the warhead compensated for the inaccuracy of delivery. The cruise missile now under development in both the United States and the Soviet Union has begun to overcome these drawbacks. It now has, or soon will have, a potential across the whole spectrum of strategic, tactical, nuclear and conventional operations. Against the existing air threat, the Soviet Union is estimated to maintain an air defence system with a total of 800,000 men. The cruise missile could be capable of outflanking the current defence system and could be produced in sufficient numbers to swamp it.

Attrition

Attrition rates are difficult to evaluate precisely because they are an amalgam of several factors, such as the rate at which replacements can be provided, the period of expected hostility, the importance of the objective under consideration, aircraft sortie rate, and the enemy's defences.

Planners have always tended to think that wars would be short and have often been proved very wrong. But never before have wars been fought with weapons of such selectivity and lethality, or with the knowledge that nuclear weapons can be called on as weapons of last resort. Whereas until the end of the 1950s it was possible to accelerate war production quickly, the increasing sophistication and complexity of military equipment has made such a move impossible, at least in the short term. In the future, wars are more likely to be fought without major prior stocking-up, at a rate and intensity foreshadowed during the 1973 Middle East war. The prospect of attrition therefore being applied to continuously dwindling force levels means that adequate stocks of men and materials must be available from the start.

Critical in interpreting attrition rates is the relationship between effort and objective. In some instances, such as the Battle of Britain or the Israeli air offensive on the Golan Heights, attrition rates were of little more than historical importance in view of the magnitude of the alternative. But even here it is difficult to establish what is an acceptable rate. The British Fighter Command's attrition of 1.57% for early July to early September 1940 almost proved unsustainable, yet later in the war 5% was thought acceptable for Bomber Command sorties. At the other end of the scale, higher attrition rates could be contemplated, particularly if the objectives were vitally important.

One reason why the apparently low attrition rate during the Battle of Britain proved nearly unsustainable was that when applied to a high sortie rate it resulted in large total losses. For similar reasons, a comparatively low attrition of 4% for Israeli air support missions during the first day of the 1973 Middle East war was considered unacceptable and, following an overnight change of tactics and an improving air situation, resulted in an overall attrition of 1.1% – although the initially higher rate probably helped achieve this. Key components were defence suppression and ECM.

Nowhere is a study of attrition more applicable than in the Central European context. The Soviet armoured offensive relies on a 'rolling bubble' of ground-based air defence systems. While some of the current generations of Soviet surface-to-air missiles (SAM) – SA-4, -6 and -8 – have a limited capability against fast, low-level targets, the systems of the next generation (SA-10) will be markedly better. Current man-portable systems have their drawbacks, whereas the ZSU-23-4 (a radar-controlled gun system) is a most potent low-level weapon. Behind this moving bubble is an integrated system of closely controlled air defence aircraft with an increasing 'look-down/shoot-down' capability, complemented by further SAM. Against this threat, studies for manned aircraft operations have shown that losses of aircraft attacking airfields in good daylight without any support operations could approach 6 out of 16 – an attrition rate of 37.5%.

Tactical routing, high-speed very low-level flight, threat-warning devices and techniques for reducing radar signatures will decrease losses, as will defence suppression by direct attack, by on-board self-screening jammers, and by stand-off ECM and decoys. Operations at night also increase protection, and, if these factors are combined, aircraft losses could be significantly reduced. Aircraft can attack the defence with stand-off missiles and with the conventionally armed cruise missile, which highlights the West's current advantage over the Soviet Union in cruise missiles and RPV technology. The weapons are small, difficult to detect and have an impressive capability against radiating targets. The emissions of radar and its associated communications networks make such targets very detectable, and it is very difficult to harden them against attack.

The growing ability to detect and engage fast, low-flying aircraft from higher-flying fighters such as the F-15 (the 'look-down/shoot-down' capability) will strengthen the defence, and it is already possible for the F-14 to engage several aircraft at a time with *Phoenix* air-to-air missiles at ranges over 160 km. But both systems depend heavily for effective use on timely identification of friend from foe, a pressing problem for all modern air forces. The introduction of airborne early warning and airborne warning and control system (AWACS) aircraft will allow defending systems to be made ready before the attack.

The Balance of Advantages

Service manuals state that air superiority in a sophisticated theatre of war is gained by a long-term battle of attrition against enemy factories making aircraft or their components, by attacking enemy aircraft on the ground and in the air, and by defending one's own air space successfully. Yet, in the short-war scenario likely in Europe, it is only the pre-emptor who could possibly achieve air superiority as it used to be known and could plan accordingly.

For many reasons airfields have tended to grow in size and complexity and to decrease in number. The positions of airfields are well catalogued, as are many of their vulnerable points. But of all the targeting options the runways are the most obvious, and their denial would have the most telling results. Very effective airfield denial weapons are currently under development – some can scatter minelets, others can penetrate concrete and explode deep in the ground to create extensive cratering. In a recent trial in the Unites States a cruise missile on a simulated counter-air mission delivered 11 out of 12 submunitions within an airfield boundary. This argues for movement away from the idea of the 'airfield fortress', both to complicate the enemy's targeting problem and to reduce dependence on long concrete runways. In West Germany, NATO offensive and defensive air forces currently operate from only 18 bases, but there are over 100 registered airstrips available which could be used.

As regards the land battle in Europe, the most critical situation is a surprise attack before the deployment of the ground forces or the preparation of their defensive positions is complete. In this situation, only air power can be brought to bear expediently to buy time for the deployment of forces and the arrival of reinforcements. A similar situation would apply in the case of a broken front, where the mobility, flexibility and concentration of air-delivered firepower can be used to plug the gap until ground forces can be redeployed. NATO concepts depend heavily on this aspect of Allied air power for their initial reaction to Warsaw Pact aggression. AWACS also have an important part to play in tracking ground movements, since their radars can reach out well beyond the forward edge of the battle area and give accurate assessments of enemy intentions (the first of 18 such aircraft is to be deployed in NATO by 1980).

While the leading elements of a ground offensive can be engaged by direct and indirect organic fire units and by the available air effort, targets beyond the range of ground weapons can only be attacked from the air. A typical Soviet front-line division consists of 800 armoured vehicles, over 200 artillery weapons and 2,000 other vehicles, 150 pieces of heavy engineering equipment and about 12,000 personnel spread over a front some 15–20 km wide and 25–50 km deep, and there is a limitation on the number of approach routes. There will also be a large number of second-echelon targets. In terms of the five classic engagement options associated with a land offensive – surprise attack, breakthrough, attack on the first and second echelons and airborne assault – the fixed-wing aircraft remains difficult to replace in its ability to penetrate to attack the back-up forces. In the case of both a surprise attack and a broken front, the air-delivered mine can be expected to be a valuable weapon for the defender in the future. Minefields can now be laid from the air and, with the technological developments which have increased the selectivity, sensitivity, compactness and lethality of the modern land mine, greater importance will be attached to this aspect of air operations.

Whereas the land battle will be marked by gradual changes, more radical developments seem likely in the sea environment.

The advent of satellite, electronic and airborne surveillance systems makes concealment very difficult, and it is already possible to keep track of major surface ship movements by satellite and aircraft. Further developments will refine the detection, identification and tracking of surface vessels and ensure that the resulting information is immediately available.

With the appearance of compact, sea-skimming, 'smart bombs' (such as maritime versions of the cruise missile), the magnitude of the threat facing surface forces has been increased quite severely. Maritime aircraft carrying these weapons are now a significant threat to shipping. Furthermore, their speed of reaction allows them to engage widely dispersed groups of vessels during a single mission. Time-on-station can be improved through new engine technology and in-flight refuelling. At the same time, maritime attack aircraft will be equipped with active and passive self-defence systems and with a limited capability to beat off defending fighters.

New Arsenals and Missions
It has become clear that the Soviet Union has moved away from the idea that an air force is an extended arm of the artillery, directly supporting the land battle, to a more Western view of air power as a crucial element in its own right in both tactical and strategic operations. As a consequence she has introduced, and continues to deploy, more complex aircraft designed to deliver heavier weapon loads over longer ranges and with greater accuracy than ever before. Soviet air doctrine in the conventional phase of a war with NATO now seems to have two aims: the early destruction of NATO's theatre nuclear capability and the maximum reduction of Allied air capability.

The new aircraft, mostly variable-geometry ('swing-wing') designs, are well suited to meet these objectives. In the initial phases of a Central Region conflict, battlefield air support would be subordinate to the deeper-ranging roles of anti-nuclear operations, air defence suppression, counter-air, and disruption of command and control systems. Nevertheless, a proportion of the ground-attack aircraft would be allocated to battlefield air

support, together with armed helicopters designed to complement the fixed-wing aircraft. Thus, over the past ten years, the Soviet Union has redefined the role of air power and produced the equipment to meet the new concept of operation.

Whereas the United States used to be ahead in nearly all areas of aviation strategy, the position is now more finely balanced, with the American lead being primarily in areas such as ECM, airborne radars and missiles, strategic and tactical transports, and in high-technology engines, airframes and weapon-aiming systems. The United States has inevitably been influenced by the Vietnam experience; she is being forced to adopt tactics of fast, low-level operations to get under Soviet air defences. But there seems to be confidence that it will be possible to penetrate.

One of the significant changes that has taken place in the last two decades is the way that air power is manoeuvred and projected in the Third World. The Soviet Union has both used proxy air forces and also flown in replacement forces and equipment, as in the war in the Middle East in 1973 – a move which the United States matched. The application of air power from a very long range can upset the local balance of power dramatically. There is also an important psychological effect attached to the use of air power, which the Rhodesian raids on Lusaka and the recent stationing of MiG-23 aircraft in Cuba illustrate. Air superiority, although no longer relevant to the Central European context, still has relevance to air operations in the Third World, where the conditions and the scale of activity will be very different.

The manned aircraft would seem likely to retain the flexibility and adaptability which will enable it to operate effectively against fast-moving forces in a way which more inert systems cannot do. However, it will become complementary to cruise and other missiles and RPV, with the aircraft becoming the lifter, carrier and positioner of weapons of increasing stand-off capability. Agility and intelligence will be built into the weapon rather than the aircraft. Since the new weapons enable the launch platform to stay some distance from enemy defences, its survivability is enhanced.

127

Index